T0134652

Sustained Simulation Performance 2017

Michael M. Resch • Wolfgang Bez • Erich Focht •
Michael Gienger • Hiroaki Kobayashi
Editors

# Sustained Simulation Performance 2017

## Proceedings of the Joint Workshop on Sustained Simulation Performance, University of Stuttgart (HLRS) and Tohoku University, 2017

*Editors*
Michael M. Resch
High Performance Computing
Center (HLRS)
University of Stuttgart
Stuttgart, Germany

Wolfgang Bez
NEC High Performance Computing
Europe GmbH
Düsseldorf, Germany

Erich Focht
NEC High Performance Computing
Europe GmbH
Stuttgart, Germany

Michael Gienger
High Performance Computing
Center (HLRS)
University of Stuttgart
Stuttgart, Germany

Hiroaki Kobayashi
Cyberscience Center
Tohoku University
Sendai, Japan

---

*Figure on Front Cover:* Streamline and slice visualisation of the velocity field at peak systole in an aortic arc with aneurysm. The vortex structures in the ascending aortic arc as well as the aneurysm are visualised by the grey-coloured, transparent isosurface. The data provided by Fraunhofer MEVIS are recorded by phase contrast magnet resonance imaging and serve as the calibration and test data for the two paper contributions of Uwe Küster, Andreas Ruopp and Ralf Schneider.

---

ISBN 978-3-319-88340-3 ISBN 978-3-319-66896-3 (eBook)
DOI 10.1007/978-3-319-66896-3

Mathematics Subject Classification (2010): 65-XX, 65Exx, 65Fxx, 65Kxx, 68-XX, 68Mxx, 68Uxx, 68Wxx, 70-XX, 70Fxx, 70Gxx, 76-XX, 76Fxx, 76Mxx, 92-XX, 92Cxx



Printed on acid-free paper

This Springer imprint is published by Springer Nature
The registered company is Springer International Publishing AG
The registered company address is: Gewerbestrasse 11, 6330 Cham, Switzerland

# Preface

The field of High Performance Computing is currently undergoing a major paradigm shift. Firstly, large-scale supercomputing systems with massively improved number crunching capabilities are now available to computational scientists all over the world. At the same time, our knowledge of how to most efficiently exploit modern processors and performance achievements is growing by leaps and bounds.

However, many domains of computational science have reached a saturation point with regard to their problem size: many scientists no longer wish to solve larger problems. Instead, they aim to solve smaller problems in a shorter amount of time, so as to quickly gain knowledge and recognise potential issues early on. Yet the current architectures are much better suited to addressing large problems than they are for the more relevant smaller problem sizes.

This particular series of workshops focuses on Sustained Simulation Performance, i.e. High Performance Computing for state-of-the-art application use cases, rather than on peak performance, which is the scope of artificial problem sizes. The series of workshops was first established in 2004 under the name Teraflop Workshop and was renamed Workshop for Sustained Simulation Performance in 2012. In general, the scope of the workshop series has expanded from optimisation for vector computers only to future challenges, productivity, and exploitation of current and future High Performance Computing systems.

This book presents the combined outcomes of the 24th and 25th workshops in the series. The 24th workshop was held at the High Performance Computing Center, Stuttgart, Germany, in December 2016. The subsequent 25th workshop was held in March 2017 at the Cyberscience Center, Tohoku University, Japan. The topics studied by the contributed papers include developing novel system management concepts (Part I), leveraging innovative mathematical methods and approaches (Part II), applying optimisation as well as vectorisation techniques (Part III), implementing Computational Fluid Dynamics applications (Part IV), and finally, exploiting High Performance Data Analytics (Part V).

We would like to thank all the contributors and organisers of this book and the Sustained Simulation Performance project. We especially thank Prof. Hiroaki Kobayashi for the close collaboration over the past years and look forward to intensifying our cooperation in the future.

Stuttgart, Germany
July 2017

Michael M. Resch
Michael Gienger

# Contents

# Part I
# System Management

# Theory and Practice of Efficient Supercomputer Management

**Vadim Voevodin**

**Abstract** The efficiency of using modern supercomputer systems is very low due to their high complexity. It is getting harder to control the state of supercomputer, but the cost of low efficiency can be very significant. In order to solve this issue, software for efficient supercomputer management is needed. This paper describes a set of tools being developed in Research Computing Center of Lomonosov Moscow State University (RCC MSU) that is intended to provide a holistic approach to efficiency analysis from different points of view. Efficiency of particular user applications and whole supercomputer job flow, efficiency of computational resources utilization, supercomputer reliability, HPC facility management—all these questions are being studied by the described tools.

## 1 Introduction

Modern supercomputing system consists of a huge amount of different software and hardware components: compute nodes, network, storage, system software tools, software packages, etc. If we want to achieve efficient supercomputer management, we need to think about all behavior aspects of these components. How efficiently users of supercomputer center consume computational resources, what jobs they run, what projects they form, how efficiently partitions and quotas are organized, is system software configured properly—all of these (and not only these) questions need to be taken into account, otherwise the efficiency of the supercomputer usage can be significantly decreased. This means that we need to control everything happening in the supercomputer.

As the supercomputers are getting bigger and more complex, this task is getting harder and harder. This explains the fact that the efficiency of most supercomputing systems is very low. For example, the average Flops performance of one core on old MSU system called Chebyshev for 3 days is just above 3% [17]. The situation is quite the same on many other current supercomputer systems.

---

V. Voevodin (✉)
Research Computing Center of Lomonosov Moscow State University, Moscow, Russia
e-mail: vadim@parallel.ru


M.M. Resch et al. (eds.), *Sustained Simulation Performance 2017*,
DOI 10.1007/978-3-319-66896-3_1

The analysis of supercomputer efficiency is further complicated by the fact that different user groups consider efficiency in different ways. Common supercomputer users are primarily interested in solving their tasks, so they mostly think about the efficiency of their particular applications. System administrators are concerned about general usage of computational resources, so they think about the efficiency of supercomputers. In turn, the management people think more globally, so their area of interest is the efficiency of the whole supercomputer center.

The task of efficient supercomputer management is really hard to solve, but the cost of low efficiency can be very high. Here is one example. One day of Lomonosov supercomputer [18] (1.7 PFlops in peak, currently #2 in Russia) maintenance costs ~$25,000. If the job scheduler hangs, a half of the supercomputer will be idle in just 2–3 h. This means that the cost of delay with a proper reaction is very high, so we need to keep control over supercomputers.

All this explains why we need efficient supercomputer management. The next question is: what is needed to achieve it? In our opinion, there are five major directions needed to be studied. Firstly, it is necessary the collect detailed information about the current state of the supercomputer and all its components. So, a *monitoring system* is needed. Monitoring system provides a huge amount of raw data that need to be filtered out to get the valuable information about the efficiency. Intellectual *analysis* and convenient *visualization* systems are needed for that purpose. Furthermore, the efficiency of supercomputer usage directly depends on the reliability of supercomputer components. This means that all-round *control* of the correctness of system functioning is required. Also, efficient supercomputer management can be very hard without easy-to-use *work management* system for helpdesk, resource project management, hardware maintenance control, etc.

There are different existing tools that help to analyze and improve efficiency of supercomputer functioning, but they are intended to solve only one or several tasks described above. Currently there is no unified approach that allows to perform holistic analysis of the supercomputer efficiency from different points of view. In Moscow State University, we are developing a toolkit aimed to solve all of these tasks. Further in this paper, six components that form this toolkit will be described in detail.

## 2 Moscow State University HPC Toolkit for Efficiency Analysis

Figure 1 shows main components of HPC toolkit for efficiency analysis being developed in Research Computing Center of Lomonosov Moscow State University. They are interconnected and complement each other to develop a holistic approach for solving the posed task.

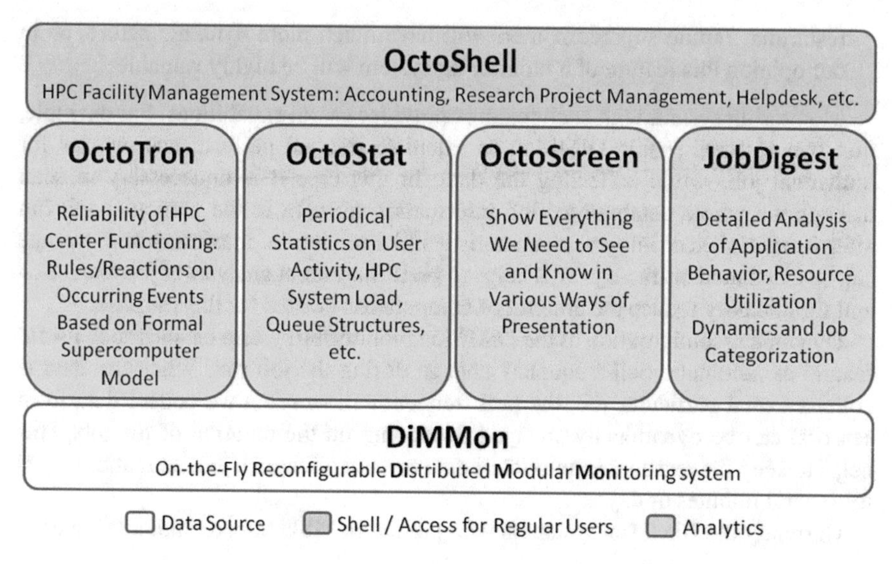


**Fig. 1** Main components of HPC toolkit for efficiency analysis developed in MSU

## *2.1 DiMMon Monitoring System*

There are many different monitoring systems that are successfully applied in practice in many supercomputer centers nowadays (Collectd [15], Nagios [4], Zenoss [20], Zabbix, Cacti, etc.). But in our opinion it will be hard to efficiently use such systems in future due to several reasons dictated by the supposed architecture of new supercomputers. Firstly, future monitoring systems need to be very scalable, up to millions of nodes. Also, they need to be easily reconfigurable, expandable and portable. And as current systems, they need to produce low overheads, but dealing with really huge amount of raw monitoring data.

Having all this in mind, Research Center in MSU started to develop DiMMon [14], new system focused mostly on performance monitoring. There are three main features that form the basis for this DiMMon approach:

1. *On-the-fly analysis*: all relevant information should be extracted from the raw data before storing to the database. This helps to greatly reduce the amount of data needed to be stored and ease further data processing.
2. *In-situ analysis*: basic processing of the monitoring data should be performed where it was collected (e.g., on a compute node), only after that it will be sent to the server side. This helps to significantly reduce the amount of data needed to be sent via communication network. Due to the fact that only simple data processing is performed locally (such as simple aggregation or speed calculation), overheads that can affect user job execution on the node are very low.
3. *Dynamic reconfiguration*: Monitoring system must be able to change its configuration (data transmission routes, collection parameters, processing rules) without

restarting. Future supercomputers will have much more dynamic nature, so in our opinion this feature of a monitoring system will be highly valuable.

Monitoring system with such features provides useful capabilities. For example, first two features enable DiMMon to calculate integral performance metrics for individual jobs while collecting the data. In this case it is unnecessary to scan through the whole database to find information relevant to the particular job run after is has finished; integral characteristics like minimum, maximum and average can be calculated on-the-fly. This helps to perform prompt analysis of job execution and significantly reduce the amount of computation needed for this purpose.

Dynamic reconfiguration of the DiMMon monitoring system enables such useful feature as automatic poll frequency change during the job run. When the data is collected on a particular job, the poll frequency (how often we collect data from sensors) can be dynamically changed depending on the duration of the job. This helps to keep the reasonable amount of data collected for each job, no matter it runs for several minutes or days.

Currently the DiMMon system is being tested in MSU Supercomputer Center.

## 2.2 *JobDigest System for Application Behavior Analysis*

JobDigest system [8] developed in MSU is intended to help users to analyze behavior dynamics of a particular job run. It processes system monitoring data on application execution such as CPU user load, number of misses in different levels of cache memory per second, number of load/store operations per second, amount of data sent/received on the node via communication network per second. All the relevant information gathered on a job run by JobDigest system is represented in form of a report that can be viewed in a web browser.

JobDigest report consists of three main blocks. First block contains some basic information received from the resource manager—job ID, user name, command used to run this job, start and finish time, number of allocated cores and nodes, etc. Second block provides integral characteristics on the job behavior. For each characteristic collected from the system monitoring data, it shows maximum, minimum and average values across all used nodes during the whole job run. This information helps to get some basic view of the general job behavior, which can be quite useful. For example, if a job runs on hundreds of nodes but does not use communication network at all (which is usually considered as undesired behavior), this will be seen in this block from the integral characteristics describing Infiniband usage intensity.

The third block contains the main information used for job behavior analysis—time series with the values of each characteristic during the application execution. An example of graphs showing these time series is presented on Fig. 2. Here CPU user load and load average (average number of processes using or waiting for CPU resources) is shown. Axis $X$ is time; axis $Y$—values of corresponding characteristic.

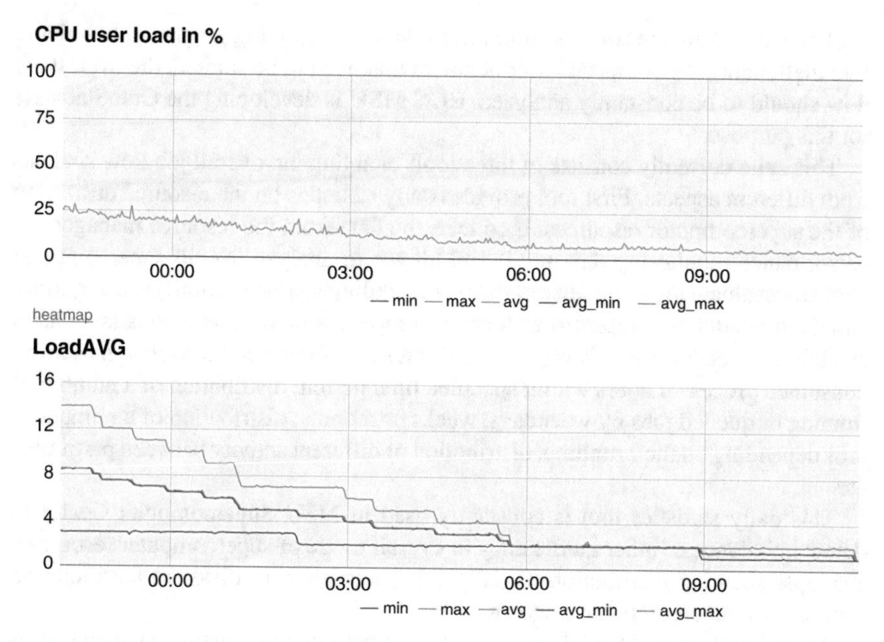


**Fig. 2** Example of JobDigest graphs with time series of two characteristics—CPU user load and load average

Different color lines represent different aggregated values—minimum, maximum and average.

These graphs show that over the time activity of this program started to slow down—both CPU user load and load average values decrease. Program activity during the last part of execution is very low meaning that a significant part of computational resources being idle. Also, the stepped nature of load average is of interest—is this behavior intentional or there is an error occurred? This behavior features cannot be detected by integral characteristics, but they can be easily found using such graphs. It should be noted that JobDigest is mainly intended to help to detect and localize performance issues in the job execution. Root cause analysis should be performed using other tools, such as profilers or trace analyzers [3, 5].

The JobDigest system is actively used in MSU Supercomputer Center on a daily basis, mostly by system administrators. The current versions of JobDigest components are freely available at Github [6].

## 2.3 *Statistics Analysis Using OctoStat*

Systems like JobDigest can be very helpful for efficiency analysis of a particular job. But such systems can be used only if we know what job run we need to analyze. And the problem is that in many cases users (sometimes even system administrators) do not know that there are some performance issues with running programs. In order to find inefficient jobs or unusual user behavior that need to be studied, the overall job flow should to be constantly analyzed. RCC MSU is developing the OctoStat suite for this purpose.

This suite currently consists of three tools enabling the overall job flow analysis from different aspects. First tool provides daily statistics on the resource utilization of the supercomputer resources. It collects the data from the resource manager (no performance monitoring data used) and allows to analyze overall supercomputer load and evaluate the optimality of queue scheduling strategies, policies and quotas. Obtained results are presented in form of a report with various graphs and charts available in web browser. Examples of statistics this report provides are: top users by consumed processor hours within specified time period; distribution of a number of running or queued jobs by weekdays, weeks or months; distribution of a number of jobs depending on their runtime; distribution of different activity between partitions; etc.

This daily statistics tool is constantly used in MSU Supercomputer Center to detect imbalance or other inefficiency in overall usage of supercomputer resources. Example showing distribution of daily job queue time in different partitions on Lomonosov supercomputer is shown on Fig. 3.

Second tool is used for job tagging. It analyzes integral dynamic characteristics (same as in JobDigest, see Sect. 2.2) of every job and automatically puts corresponding tags on it. This helps to divide jobs into different categories and easily detect suspicious or abnormal behavior. For example, tag "suspicious" is assigned to a job if average CPU user load is less than 10% and load average is less than 0.9; tag "short" is assigned to a job with execution time less than 1 h. These tags are automatically assigned not only to finished but also to running jobs, which

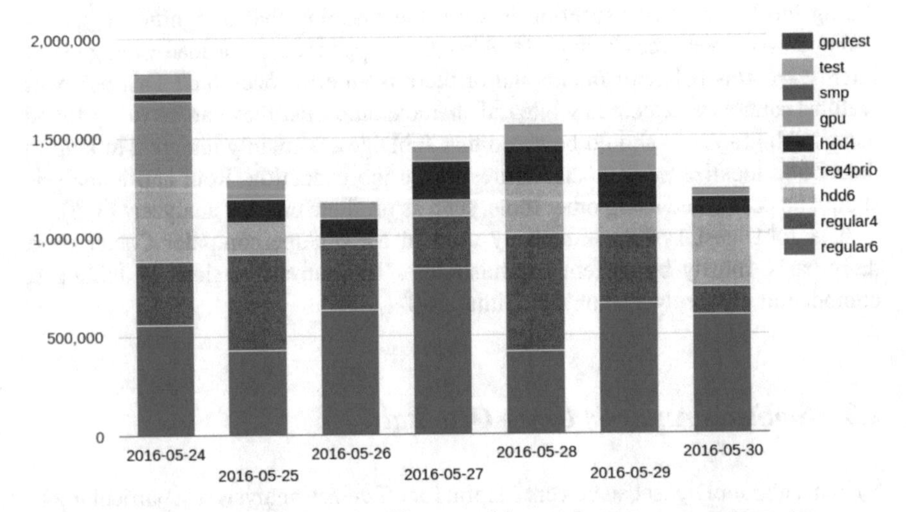


**Fig. 3** Statistics on job waiting time in different partitions on Lomonosov supercomputer

helps to promptly notify users about potential issues with current execution of their programs.

The goal of other tool in the OctoStat suite is to detect unusual behavior (anomalies) in the job flow by analyzing job efficiency [19]. This tool analyzes time series of dynamic characteristics describing the behavior of the program, using the same performance monitoring data as in JobDigest system. This tool is based on machine learning methods for detecting abnormal program execution, due to two reasons: (1) the criteria of unusual job behavior are unclear and currently cannot be precisely determined; (2) these criteria can differ significantly for different supercomputers.

The time series for a particular job is divided into a set of intervals identifying different logical parts of its execution. Each interval can belong to one of three classes: normal, abnormal or suspicious. After each interval is classified, the final classification of the entire application is performed based on the results of interval classification. An approach based on Random Forest algorithm is used for interval classification; final classification is defined by a number of criteria mostly based on a total number of processor hours consumed by intervals of each class.

Random Forest based classifier was trained on 270 normal, 70 abnormal and 180 suspicious intervals from 115 real-life applications from Lomonosov-2 supercomputer. This resulted in the accuracy of ~0.94 on the test set. Currently this classifier provides daily digest with a list of found abnormal and suspicious jobs.

## 2.4 *OctoTron: Autonomous Life of Supercomputers*

Efficient usage of a supercomputer is impossible without reliable functioning of its components. Talking about the reliability of the supercomputer, we have to deal with the following problem: in most cases nowadays we can only hope that some hardware or software component is working correctly, until there is an evidence that it has failed. But what we really need is a guarantee—if something goes wrong inside a supercomputer, we shall be notified about it immediately. We want supercomputer to behave in a way we expect it should behave. In other words, we want our expectations to match the reality.

Modern supercomputer is huge, so it is nearly impossible to manually control its state to a full extent. But supercomputer can do it itself; we only need to precisely define what "our expectations" means. This can be done the following way. The reality can be described the way it is usually done nowadays—by the means of a monitoring system. And expectations can be determined using the model of a supercomputer. This model describes all the HW/SW components in the system and their interconnections. If our expectations (model of a supercomputer) do not match the reality (system monitoring data), an emergency situation is detected.

This idea was implemented in the OctoTron system [1] developed in RCC MSU. The model used in this system is a graph; vertices are the components of the supercomputer (nodes, fans, switches, racks, network cards, HDD disks, etc.), edges—different types of connections between components. For example, an edge can be type of "contain", "chill", "power", "infiniband", etc. The more complete this graph, the more precise the functionality of the OctoTron system. The size of the model of Lomonosov supercomputer (12K CPUs, 2K GPUs, 5.8K nodes, 1.7 PFlops peak) is 116K vertices, 332K edges and 2.4M attributes. Each attribute describes one component characteristic we want to control. This could be temperature on a node, number of ECC memory errors, status of a UPS and so on.

The verification of matching reality and expectations is done by a set of formal rules. Each rule describes one particular emergency situation (i.e., mismatch between the model and monitoring data) that needs to be detected. Here are some examples of emergency situations that can be detected by the OctoTron system:

- Easy to detect emergencies:
  - hardware failures (disks, memory, ports);
  - software failures (required services are not working);
  - network failures (servers or equipment are not visible);
  - bad equipment condition (overheating, running out of space).
- More complex cases (checking a set of conditions):
  - certain ratio between the number of failed and working sensors exceeded (example: checking temperature in a hot aisle);
  - component operating modes mismatch (example: different modes on two ports of one Ethernet cable);
  - errors that depend on a composite state of components occurred (example: checking total job count in all queues).

When the emergency situation is detected and the rule is triggered, OctoTron needs to react. The most common types of reactions available are logging or notification (SMS or email sent to system administrators). Some more complex reactions can also be applied: equipment shutdown (e.g., in case of fire), bad nodes isolation (making them unavailable for users' jobs), monitoring agents restart, or execution of a custom script developed by the administrator.

It should be noted that the OctoTron system is not a replacement but rather a complement for the existing monitoring systems. The primary goal of latter ones is to collect all necessary information about the current state of the supercomputer. OctoTron uses this information to perform fine analysis of possible abnormal or incorrect behavior. Currently there are no solutions or researches similar to OctoTron approach to be found. The close one is Iaso [7] system by NUDT University—this system can automatically detect complex emergency situations and find root causes, but it is not publicly available at the moment.

OctoTron system is actively used on Lomonosov system and is being implemented on Lomonosov-2 supercomputer. It is available under open MIT license [12].

## 2.5 *OctoScreen Visualization System*

Collecting data using monitoring systems and studying this data using different analysis tools is a very important step to achieve efficient supercomputer management. But in many cases it is difficult to perceive obtained results without their proper representation. That is the reason RCC MSU is developing the OctoScreen visualization suite [10]. This system is intended to provide handy visualization of supercomputer-related data for every need. It is aimed to help all supercomputer groups (common users, system administrators and management) and can be used for many different purposes—short digest with the most important data about supercomputer center functioning, monitoring data representation helping to control current supercomputer state, lecture screen, etc.

Figure 4 shows one of such examples for Lomonosov supercomputer—a screen that can be used in lectures or excursions to show the main interesting information about the supercomputer. In the left part, each graph is a timeline showing the dynamics of different characteristics change within last several days. For example, it can be seen that the number of available and used CPU cores almost does not change as well as the average temperature in both cold and hot aisles. But the number of queued and running tasks varies significantly. There are two graphs in the right part; one shows the map with geographical distribution of supercomputer users, the other shows the current temperature distribution in the computer room.

OctoScreen is used in practice for other purposes as well. A mobile web-site is developed to provide the most important basic information on the current state of

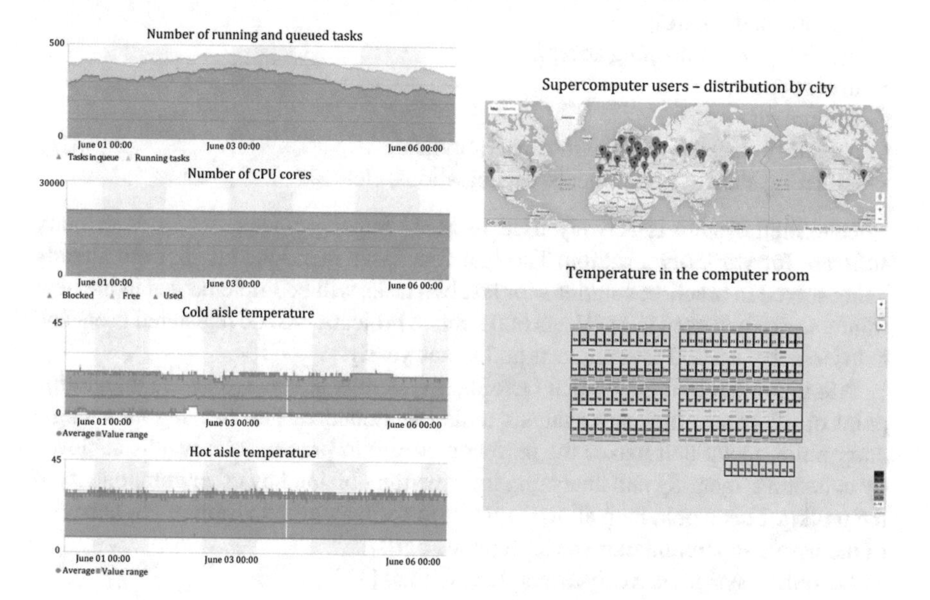


**Fig. 4** Example of visualization screen available in the OctoScreen system

supercomputer that helps management to quickly determine if the jobs are running normally. Another useful example—hardware state monitoring screen available online for system administrators that tells when the hardware error occurred and how long it took to fix this error.

## 2.6 *OctoShell System for Work Organization*

System administrators of many big supercomputer centers have to deal with a difficult task of maintaining a huge stack of support software. User management tools (LDAP, NIS, batch systems like Slurm [13] or Torque [16], etc.), hardware monitoring systems and services, user support software like ticket management systems—all this usually independent software need to be in a working and consistent state. Most of this work is poorly automated, which results in various issues: big number of consistency errors, significant processing time and poor portability. There are few existing systems (e.g. Bright Cluster Manager [2]) that provide needed solutions, but all of them are not open source.

The OctoShell system [9] is intended to help solving these issues. It provides single entry point for many aforementioned services, which ease the automation process and helps to make the solution easily portable and extendable. This system is used as a front-end by all supercomputer groups and is intended to solve the following tasks:

- project management, which includes resource allocation as well as user account creation and control;
- providing and managing access;
- user support;
- equipment service;
- inventory tracking of all hardware equipment;
- gathering statistics on projects, organizations, etc.

OctoShell system is actively used in MSU Supercomputer Center as primary software for work organization. The first four tasks mentioned earlier are already being solved in practice; solutions for last two tasks will be implemented in the near future. OctoShell used in MSU has 600+ active projects, ~2700 registered users and helps to solve ~1000 user requests and issues yearly.

It is important to mention that OctoShell system is planned to be the integration point of all other efficiency analysis tools mentioned earlier. Serving as a single entry point, OctoShell makes the perfect platform to present joint results achieved by collecting, merging and analyzing information obtained by other components of the toolkit. This would help all user groups to form a unified vision of the behavior of the whole supercomputer and its separate parts.

OctoShell system is freely available at Github [11].

## 3 Conclusion

This paper describes the toolkit for efficient supercomputer management being developed in Research Computing Center of Lomonosov Moscow State University. This toolkit consists of six components aimed at analyzing different aspects of supercomputer behavior, which helps to get a deep understanding of supercomputer functioning efficiency. DiMMon is a dynamically reconfigurable monitoring system that is capable of on-the-fly and in-situ data processing. JobDigest system provides detailed reports on the dynamic behavior of particular applications based on performance monitoring data. OctoStat analyzes the overall job flow to discover imbalance in computational resource utilization and detect jobs with unusual and inefficient behavior, which can be analyzed in more detail using tools like JobDigest. OctoTron is a system for ensuring supercomputer reliability which is based on formal model of a supercomputer. It allows to automatically detect different types of emergency situations and promptly perform corresponding reactions. OctoScreen is aimed at handy visualization of any kind of information about the current state of supercomputer and its components for every need. Finally, OctoShell is used as a single point for work organization within supercomputer center that helps to ease and automate such tasks as user support, project management, equipment service, etc. All these tools complement each other to develop a unified approach for efficiency analysis that helps to achieve the goal of efficient supercomputer management.

These systems being developed are open source and planned to be fully portable and available to the supercomputer community.

**Acknowledgements** This material is based upon work supported in part by the Russian Found for Basic Research (grant No. 16-07-00972) and Russian Presidential study grant (SP-1981.2016.5).

## References

1. Antonov, A., Nikitenko, D., Shvets, P., Sobolev, S., Stefanov, K., Voevodin, V., Voevodin, V., Zhumatiy, S.: An approach for ensuring reliable functioning of a supercomputer based on a formal model. In: Parallel Processing and Applied Mathematics: 11th International Conference, PPAM 2015, Krakow, September 6–9, 2015. Revised Selected Papers, Part I, pp. 12–22. Springer, Cham (2016). doi:10.1007/978-3-319-32149-3_2. http://link.springer.com/10.1007/978-3-319-32149-3_2
2. Bright Cluster Manager home page. http://www.brightcomputing.com/product-offerings/bright-cluster-manager-for-hpc. Cited 15-06-2017
3. Geimer, M., Wolf, F., Wylie, B.J.N., Ibrahim, E., Becker, D., Mohr, B.: The Scalasca performance toolset architecture. Concurr. Comput. Pract. Experience **22**(6), 702–719 (2010). doi:10.1002/cpe.1556. http://doi.wiley.com/10.1002/cpe.1556
4. Infrastructure Monitoring System Nagios. https://www.nagios.org/. Cited 15 Jun 2017
5. Jagode, H., Dongarra, J., Alam, S., Vetter, J., Spear, W., Malony, A.D.: A holistic approach for performance measurement and analysis for petascale applications. In: Computational Science – ICCS 2009, pp. 686–695. Springer, Berlin (2009). doi:10.1007/978-3-642-01973-9_77. http://link.springer.com/10.1007/978-3-642-01973-9_77

6. JobDigest Components. https://github.com/srcc-msu/job_statistics. Cited 15 Jun 2017
7. Lu, K., Wang, X., Li, G., Wang, R., Chi, W., Liu, Y., Tang, H., Feng, H., Gao, Y.: Iaso: an autonomous fault-tolerant management system for supercomputers. Front. Comp. Sci. **8**(3), 378–390 (2014). doi:10.1007/s11704-014-3503-1. http://link.springer.com/10.1007/s11704-014-3503-1
8. Mohr, B., Voevodin, V., Gimenez, J., Hagersten, E., Knupfer, A., Nikitenko, D.A., Nilsson, M., Servat, H., Shah, A., Winkler, F., Wolf, F., Zhukov, I.: The HOPSA workflow and tools. In: Tools for High Performance Computing 2012, pp. 127–146. Springer, Berlin (2013). doi:10.1007/978-3-642-37349-7_9. http://link.springer.com/10.1007/978-3-642-37349-7_9
9. Nikitenko, D.A., Voevodin, V.V., Zhumatiy, S.A.: Octoshell: large supercomputer complex administration system. Bull. South Ural State Univ. Ser. Comput. Math. Softw. Eng. **5**(3), 76–95 (2016). doi:10.14529/cmse160306. http://vestnik.susu.ru/cmi/article/view/3998
10. Nikitenko, D.A., Zhumatiy, S.A., Shvets, P.A.: Making large-scale systems observable - another inescapable step towards exascale. Supercomput. Front. Innov. **3**(2), 72–79 (2016). doi:10.14529/jsfi160205. http://superfri.org/superfri/article/view/96
11. OctoShell Source Code. https://github.com/%5Cshell/%5Cshell-v2. Cited 15 Jun 2017
12. OctoTron Framework Source Code. https://github.com/srcc-msu/OctoTron. Cited 15 Jun 2017
13. Slurm Workload Manager. https://slurm.schedmd.com/. Cited 15 Jun 2017
14. Stefanov, K., Voevodin, V., Zhumatiy, S., Voevodin, V.: Dynamically reconfigurable distributed modular monitoring system for supercomputers (DiMMon). Proc. Comput. Sci. **66**, 625–634 (2015). doi:10.1016/j.procs.2015.11.071. http://linkinghub.elsevier.com/retrieve/pii/S1877050915034201
15. System Statistics Collection Daemon Collectd. https://collectd.org/. Cited 15 Jun 2017
16. TORQUE Resource Manager. http://www.adaptivecomputing.com/products/open-source/torque/. Cited 15 Jun 2017
17. Voevodin, V., Voevodin, V.: Software system stack for efficiency of exascale supercomputer centers. Technical Report (2015)
18. Voevodin, V., Zhumatiy, S., Sobolev, S., Antonov, A., Bryzgalov, P., Nikitenko, D., Stefanov, K., Voevodin, V.: The practice of “Lomonosov” supercomputer. Open Syst. DBMS **7**, 36–39 (2012)
19. Voevodin, V., Voevodin, V., Shaikhislamov, D., Nikitenko, D.: Data mining method for anomaly detection in the supercomputer task flow. In: Numerical Computations: Theory and Algorithms, The 2nd International Conference and Summer School, pp. 090015-1–090015-4. Pizzo Calabro (2016). doi:10.1063/1.4965379. http://aip.scitation.org/doi/abs/10.1063/1.4965379
20. Zenoss – Monitoring and Analytics Software. https://community.zenoss.com/home. Cited 15 Jun 2017

# Towards A Software Defined Secure Data Staging Mechanism

**Susumu Date, Takashi Yoshikawa, Kazunori Nozaki, Yasuhiro Watashiba, Yoshiyuki Kido, Masahiko Takahashi, Masaya Muraki, and Shinji Shimojo**

**Abstract** Recently, the necessity and importance of supercomputing has been rapidly increasing in all scientific fields. Supercomputing centers in universities are assumed to satisfy scientists' diverse demands and needs for supercomputing. In reality, however, medical and dental scientists who treat security-sensitive data have difficulties using any supercomputing system at a supercomputing center due to data security. In this paper, we report on our on-going research work towards the realization of a supercomputing environment where separation and isolation of our supercomputing environment is flexibly accomplished with Express Ethernet technology. Specifically, in this paper, we focus on an on-demand secure data

---

S. Date (✉) • Y. Kido • S. Shimojo
Cybermedia Center, Osaka University, 5-1 Mihogaoka, Ibaraki, Osaka 567-0047, Japan
e-mail: date@cmc.osaka-u.ac.jp; kido@cmc.osaka-u.ac.jp; shimojo@cmc.osaka-u.ac.jp

T. Yoshikawa
System Platform Research Laboratories, NEC Corporation, 1753 Shimonumabe, Nakahara-ku, Kawasaki, Kanagawa 211-8666, Japan

Cybermedia Center, Osaka University, 5-1 Mihogaoka, Ibaraki, Osaka 567-0047, Japan
e-mail: yoshikawa@cd.jp.nec.com; tyoshikawa@cmc.osaka-u.ac.jp

K. Nozaki
Division of Medical Informatics, Osaka University Dental Hospital, 1-8 Yamadaoka, Suita, Osaka 565-0871, Japan
e-mail: knozaki@dent.osaka-u.ac.jp

Y. Watashiba
Graduate School of Information Science, Nara Institute of Science and Technology, 8916-5 Takayama, Ikoma, Nara 630-0192, Japan
e-mail: watashiba@is.naist.jp

M. Takahashi
System Platform Research Laboratories, NEC Corporation, 1753 Shimonumabe, Nakahara-ku, Kawasaki, Kanagawa 211-8666, Japan
e-mail: m-takahashi@ex.jp.nec.com

M. Muraki
Strategic Technology Center, TIS Inc., 17-1, Nishishinjuku 8-chome, Shinjuku-ku, Tokyo 160-0023, Japan
e-mail: muraki.masaya@tis.co.jp


M.M. Resch et al. (eds.), *Sustained Simulation Performance 2017*,
DOI 10.1007/978-3-319-66896-3_2

staging mechanism that interacts with a job management system and a software defined networking technology, thus enabling minimum data exposure to third parties.

## 1 Introduction

The necessity and importance of supercomputing has been rapidly rising in all scientific fields. This rising necessity and importance can be explained from the following four factors. First, the recent advance in scientific data measurement devices has allowed scientists to obtain high resolution scientific data in both temporally and spatially, and this has resulted in an increasing amount of data obtained with such devices. For example, according to [16], the Large Synoptic Survey Telescope (LSST) is expected to generate 15 TB a day. Also, approximately 15 PB of experimental data is annually generated and processed at the Large Hadron Collider (LHC), an experimental facility for high energy physics [2]. Second, the development of current processors and accelerators has advanced dramatically. Although discussions about 'Post Moore's era' are more and more frequent these days, new-generation processors and accelerators have been continuously researched and released. The Intel Xeon Phi (Knight Landing) processor and the NVIDIA GPU accelerator (Pascal and Volta) are representative examples of such processors and accelerators. NEC's future vector processor is another example of a cutting-edge processor [11]. These cutting-edge processors and accelerators have allowed researchers to perform an in-depth and careful analysis of scientific data observed and measured through the use of scientific measurement devices on supercomputing systems with such processors and accelerators. Third, networking technologies have advanced greatly. Today, 10 Gbps, and even 100 Gbps-class networks are available as world-scale testbeds for scientific research [4, 14, 15]. This advancement means that scientists and researchers can move scientific data more quickly in a research collaboration environment where research institutions, universities, and industries are connected. Furthermore, the potential and feasibility of data movement using advanced networking technology benefits the aggregation and sharing of scientific knowledge and expertise for problem solving. Finally, the needs and demands of high-performance data analysis (HPDA) have led to increased demands on supercomputing. Scientists from every field are enthusiastic about applying computationally intensive artificial intelligence (AI) technologies that exemplify deep learning in their own scientific domains.

Despite the increasing demand on supercomputing, however, the Cybermedia Center (CMC) [9], a supercomputing center at Osaka University in Japan, is facing two serious problems because of the recent diversification of users' requests and requirements from their high-performance computing environments, in addition to the strong user requests and demands for larger computational power. The first problem is the low utilization of supercomputing systems due to an inflexibility in resource configuration, and the second problem is the loss of supercomputing opportunities due to data security. These two problems hinder the efficient and effective use of supercomputing systems for scientific research.

The first problem can be explained from the many choices scientists have today in terms of the hardware and software for acceleration of their programs and thus, each scientist tends to have his/her own favorite supercomputing environment. For example, some scientists may want to use OpenMP to achieve a high degree of thread-level parallelism on a single computing node equipped with two 18-core processors, while others may want to use MPI to achieve inter-node parallelism on 16 computing nodes. Also, a scientist may want to perform GROMACS [1] on a single computing node with four GPUs, while others may want to perform LAMMPS [8] on a single computing node with two GPUs. Taking these users' diverse requests and requirements into consideration, supercomputing centers like the CMC should have a flexibility in resources to accommodate the diversity and heterogeneity of user requests pertaining to supercomputing systems. Based on this observation and insight described above, our research team has been working on the research and development of a more flexible supercomputing system. We have published the results and achievements obtained so far in [3, 10] and therefore do not present how we have approached this problem in this paper.

The second problem for scientists is how to treat a large amount of privacy-rich and confidential data, especially in the medical and dental sciences, such as is the case at Osaka University. Medical and dental scientists acquire a lot of such privacy-rich and confidential data which they want to analyze with a supercomputer system for the prediction of patients' future disease risks, for assessment of the severity of the patient's case, and for the understanding of brain function, etc. All of these situation require deep learning techniques on a high-performance simulation. Unfortunately, in reality, medical and dental scientists have great difficulties using the supercomputing systems at the CMC because of privacy issues. To assist scientific researches treating this type of security-sensitive scientific data, a technical solution that enables security-sensitive data to be treated is essential from the point of a supercomputing service provider.

In this paper, we present and report the research work in progress for the second problem. More specifically, an on-demand secure data staging mechanism that enables minimum exposure of security-sensitive scientific data, which we have envisaged and been prototyping, is overviewed. Technically, the mechanism leverages Software Defined Networking in cooperation with a job management system for the mechanism. The organization of this paper is as follows. Section 2 briefly overviews the challenge and issues in this early stage of our research. Next, we describe our basic idea to approach this challenge and then we explain the key technologies composing the mechanism in Sect. 3. Subsequently, in Sect. 4, the overview of the mechanism is shown. Section 5 concludes this paper.

## 2 Challenges and Issues

To tackle data security, we have started a discussion about data security issues with the Division of Medical Informatics at the Osaka University Dental Hospital so that dental scientists can utilize supercomputing systems for their own research. The biggest hurdle and problems to overcome so far are the regulations and guidelines set forth by the Ministry of Health, Labour and Welfare, Japan, which has strictly required organizations and/or scientists that treat privacy-rich data in adherence with the government's regulations and Ministry guidelines. The Division of Medical Informatics at the Osaka University Dental Hospital has set up their own security policy and rules based on their regulations and guidelines for managing controlling their security-sensitive data.

The second hurdle and problem is how we can make a supercomputing environment to include the network between at the Division of Medical Informatics at the Osaka University Dental Hospital and the CMC dedicated only to dental scientists who are willing to use supercomputing systems at the CMC. As described above, the CMC is in charge of delivering a supercomputing environment to researchers in universities and research institutions. Thus, the supercomputing systems are inherently expected to be shared by many scientists and researchers at the same time. Dental and medical scientists, however, do not want to share data and need to securely move data from the hospital to the CMC's supercomputing environment, perform their computation on a supercomputing environment dedicated to them, and move their data and computational results back to their departments. In other words, we, at the CMC, need to find a way to service the privacy of data from the dental hospital.

## 3 Key Technologies

Our approach to the second problem described in the previous section synergically makes use of three key technologies: Software Defined Networking (SDN), Express Ethernet (ExpEther) technology, and Job Management System to realize a secure and isolated supercomputing environment where dental scientists can perform their own computations. The following subsections explain these three key technologies.

### *3.1 Software Defined Networking*

Software Defined Networking [12, 17] is a new concept of the construction and management of computer networks. Traditional computer networking facilities have a built-in implementation of network protocols such as Spanning Tree Protocol (802.1D) and Tagged VLAN (802.1Q). SDN provides a systematic separation of two essential functionalities of such networking facilities, namely, data forwarding and decision making. In SDN, the data forwarding part is called the *Data Plane* and the other part, which decides how each data should be forwarded in a network and conveys the decision to appropriate networking facilities, is called the *Control Plane*. Control Plane is usually implemented as a software program, hence, the name Software-Defined.

Separating the Data Plane and the Control Plane can deliver many benefits to those who construct and manage computer networks. Most significantly, the separation of the Control Plane from physical networking facilities such as Ethernet switches makes replacing protocol handling modules possible. This can be done quickly by replacing the software program installed in the Control Plane, without updating any firmware or configurations on the actual networking facilities. This feature is beneficial for operators who want to realize their own automated network management suited to their particular businesses, or researchers who want to try their new networking protocols or traffic management scheme. Other benefits may include high-level interoperability between the Data Plane and the Control Plane, and applicability of software engineering techniques when developing new networking protocols.

## 3.2 ExpEther Technology

Express Ethernet (ExpEther) technology basically virtualizes PCI Express over the Ethernet [5] and creates a single hop PCI Express switch even if the Ethernet network is composed of multiple switches. A promising feature of this architecture is that we can put as many computers and devices as necessary in a single Ethernet network without limits to the connection distance.

Another feature of ExpEther technology is that it allows us to attach and detach such devices to and from computers in a software-defined manner. This feature utilizes the characteristics of PCI in that its configuration is automatically executed among ExpEther chips that have the same group ID. In other words, by controlling the group ID, the computer hardware can be reconfigured.

## 3.3 Job Management System

The Job Management System is usually deployed to a high-performance cluster system for the purpose of load balancing and throughput improvement. This system is in charge of receiving resource requests from users, scheduling assignments of processor resources and then assigning an appropriate set of resources to each resource request based on its scheduling plan. Numerous job management systems have been proposed and implemented. Examples of such job management system include PBS [6], NQS [7] and Open Grid Scheduler/Grid Engine (OGS/GE) [13].

In general, the job management systems are also known as queuing systems, and these job management systems use multiple queues categorized by resource limitations and requests' priorities for resource allocation of the system. Thus, the job management systems completely understand when and which job should start. In this research, our job management system interacts with SDN and ExpEther technology to realize a computing environment dedicated to the job owner.

## 4 Proposal

Figure 1 shows an overview of our envisioned supercomputing environment. As illustrated in the figure, the job management system receives user's requests in the form of a batch script. The received job requests are stored in a queue in the job management system and then wait to be dispatched. At this time the contents of the job requests are parsed and understood by the job management system in terms of what kind of devices, for ex., GPU and SSD (Solid State Drive) in a resource pool connected to a ExpEther network are required. Based on the information on resources requested by jobs in the queue and the usage information of computing nodes, the job management system attempts to coordinate an optimal allocation plan of resources to each job request, taking into account the availability of processors and devices in the resource pool. When a certain job request's turn comes, the job management system interacts with the ExpEther manager to prepare for a set of computing nodes reconfigured with user-requested devices in a resource pool through the use of OpenStack Ironic, which is the module for bare metal machine management functionality. Figure 2 diagrams the inside interaction of the resource management software. The above-mentioned interaction mechanism between the job management system and ExpEther technology allows users to specifically request their own desired computing environment. For example, two computing nodes, each of which has four GPU nodes, can be specified through the batch script file to the job management system. The details of this mechanism have already been reported in [10].

For the security issue described in Sect. 2, our envisioned supercomputing environment plans to take advantage of network programmability brought by SDN in cooperation with the job management system. As described in the above paragraph, the job management system can learn when a certain job is run on which computing nodes. In our envisioned computing environment, we design the data stage-in and stage-out functionality so that the connectivity of a network between data storage where security-sensitive data is located and SSD to be connected to computing nodes via the ExpEther network is guaranteed only before and after the job treating the security-sensitive data is executed. As explained in Sect. 3.1, SDN enables the control of packet flows in a software programming manner. This means that the on-demand control of network connectivity can be achieved in response to the necessity of data movement. This prominent feature of SDN is considered promising in lowering the risk of data exposure to third parties. In our envisioned

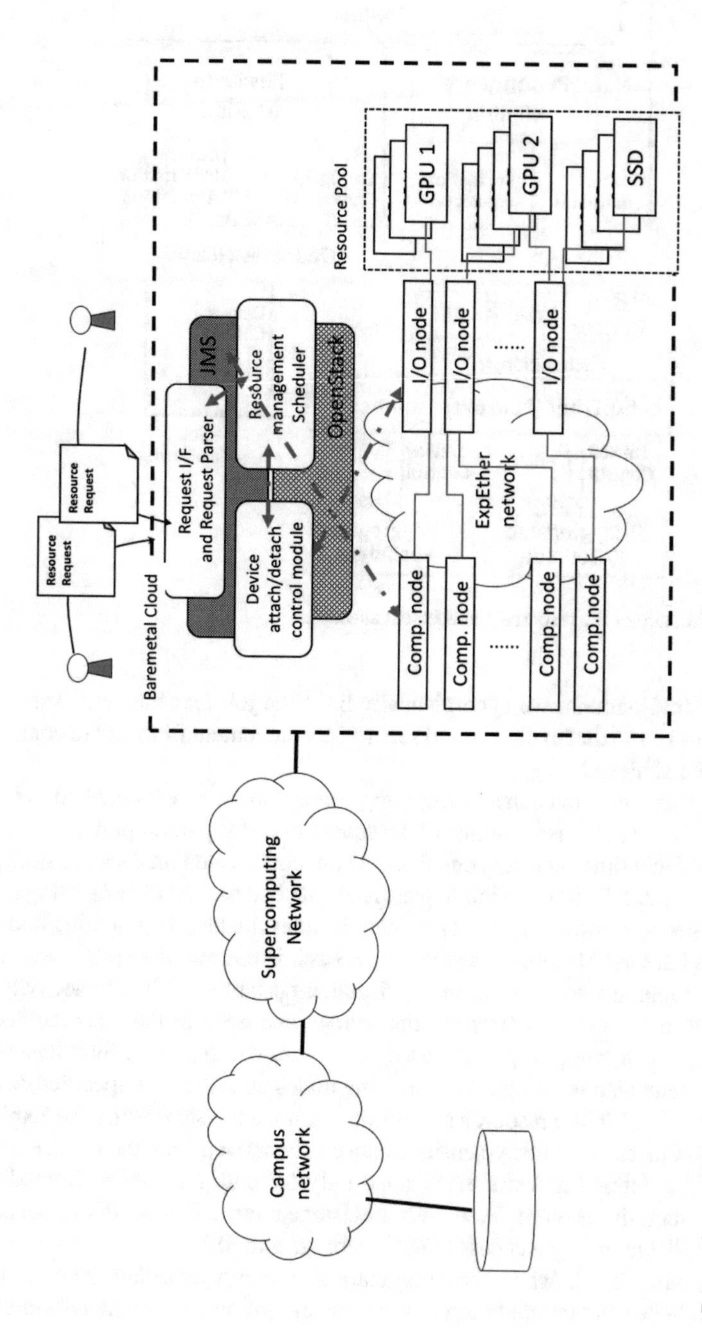


**Fig. 1** Proposal overview

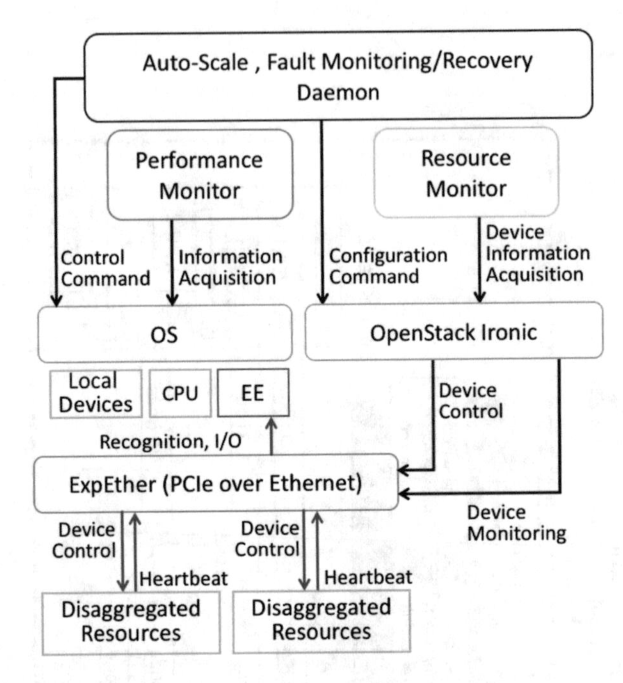


**Fig. 2** Architecture of the resource management scheduler

computing environment, we synergistically have the job management system and SDN interlocked so that exclusive and secure data movement from and to computing nodes can be achieved.

In our plan, the envisioned computing environment is expected to work as follows. A dental scientist submits a job request to our supercomputing systems at the CMC. At this time security-sensitive data are still located on a secure storage in the dental hospital. Before the job request is dispatched to a set of computing nodes, the envisioned supercomputing environment isolates the target computing nodes by preparing a bare metal environment on OpenStack Ironic and then preventing other users from logging into it. Next, the environment establishes the connectivity and reachability of the network between the storage and SSD in the resource pool to be connected to a computing node used for secure computation. Simultaneously, the environment also reconfigures computing nodes with the user-specified devices including the SSD in the resource pool of our computer system using the ExpEther technology. Currently, security-sensitive data can be moved from the storage to SSD on a software-defined network established only for data movement. Immediately after data stage-in is completed, our envisioned environment disconnects the network so that no one can access both the storage and SSD.

On the other hand, when the computation is completed, data staging-out is performed. When the computation is finished, the job management system establishes the network between the storage and the SSD again and then move the computational results back to the storage. After finishing the data movement, the staged-in data and computational results are completely removed and then SSD is detached from the computing node so that no one can access SSD.

In this research we aim to reduce the risk of data exposure to third parties by realizing the above-mentioned computing environment. At the time of writing this paper, we have been working on the prototype of the data stage-in and stage-out mechanism interlocked with the job management system and SDN. Furthermore, FlowSieve [18], a network access control mechanism leveraging SDN, which we have prototyped, can be applied in the future for enhancement of data security.

## 5 Conclusion

This paper has reported the research in progress towards the realization of an on-demand secure data staging mechanism that enables minimum exposure of security-sensitive data to third parties. Currently, we have been working on the prototype of the mechanism and the integration of it with SDN into JMS, in the hope that the achievement of this research will enable scientists to analyze such data on supercomputing systems at the CMC. At the same time, through continuing collaboration with scientists who need to treat security-sensitive data, we have recognized that there are still many unsolved security issues related to the regulations and guidelines mentioned in Sect. 2.

**Acknowledgements** This work was supported by JSPS KAKENHI Grant Number JP16H02802 and JP26330145. This research achievement is partly brought through the use of the supercomputer PC cluster for large-scale visualization (VCC).

## References

1. Abraham, M.J., Murtola, T., Schulz, R., Pall, S., Smith, J.C., Hess, B., Lindahl, E.: GROMACS: high performance molecular simulations through multi-level parallelism from laptops to supercomputers. SoftwareX **1–2**, 19–25 (2015)
2. Bird, I.: Computing for the Large Hadron Collider. Annu. Rev. Nucl. Part. Sci. **61**(1), 99–118 (2011). doi:10.1146/annurevnucl-102010-130059
3. Date, S., Kido, Y., Khureltulga, D., Takahashi, K., Shimojo, S.: Toward flexible supercomputing and visualization system. Sustained Simulation Performance 2015, pp. 77–93. Springer, Cham (2015). doi:10.1007/978-3-319-20340-9_7
4. ESnet. https://www.es.net/
5. ExpEther (Express Ethernet) Consortium. http://www.expether.org/
6. Henderson, R.L.: Job scheduling under the portable batch system. In: Job Scheduling Strategies for Parallel Processing, vol. 949, pp. 279–294. Springer, Cham (1995)
7. Kingsbury, B.A.: The network queuing system. Technical Report, Sterling Software (1992)
8. LAMMPS Molecular Dynamics Simulator. http://lammps.sandia.gov/
9. Large-Scale Computer System, The Cybermedia Center at Osaka University. http://www.hpc.cmc.osaka-u.ac.jp/en/

10. Misawa, A., Date, S., Takahashi, K., Yoshikawa, T., Takahashi, M., Kan, M., Watashiba, Y., Kido, Y., Lee, C., Shimojo, S.: Highly reconfigurable computing platform for high performance computing infrastructure as a service: Hi-IaaS. In: The 7th International Conference on Cloud Computing and Services Science (CLOSER 2017), April 2017, pp. 135–146. doi:10.5220/0006302501630174
11. Momose, S.: NEC supercomputer: its present and future. In: Sustained Simulation Performance 2015, pp. 95–105. Springer, Cham (2015). doi:10.1007/978-3-319-20340-9_8
12. Nunes, B.A., Mendonca, M., Nguyen, X.N., Obraczka, K., Turletti, T.: A survey of software-defined networking: past, present, and future of programmable networks. IEEE Commun. Surv. Tutorials **16**(3), 1617–1634 (2014)
13. Open Grid Scheduler: The Official Open Source Grid Engine. http://gridscheduler.sourceforge.net/.
14. Science Information Network 5 (SINET5). http://www.sinet.ad.jp/en/top-en/
15. Singapore Advanced Research and Education Network (SingAREN). https://www.singaren.net.sg/
16. The Large Synoptic Survey Telescope. http://www.lsst.org/
17. Xia, W., Wen, Y., Foh, C.H., Niyato, D., Xie, H.: A survey on software-defined networking. IEEE Commun. Surv. Tutorials **17**(1), 27–51 (2015)
18. Yamada, T., Takahashi, K., Muraki, M., Date, S., Shimojo, S.: Network access control towards fully-controlled cloud infrastructure. Ph.D. Consortium. In: 8th IEEE International Conference on Cloud Computing Technology and Science (CloudCom2016), December 2016. doi:10.1109/CloudCom.2016.0076

# Part II
# Mathematical Methods and Approaches

# The Numerical Approximation of Koopman Modes of a Nonlinear Operator Along a Trajectory

**Uwe Küster, Ralf Schneider, and Andreas Ruopp**

**Abstract** The spectral theory of linear operators enables the analysis of their properties on stable subspaces. The Koopman operator allows to extend these approaches to a large class of nonlinear operators in a surprising way. This is even applicable for numerical analysis of time dependent data of simulations and measurements. We give here some remarks on the numerical approach, link it to spectral analysis by the Herglotz-Bochner theorem and are doing some steps for significance for partial differential equations.

## 1 Introduction

This paper is directly related to a first part [7] and a second part [8] from the author and is to be considered as an extension of the numerical approaches.

Linear operators are used and deeply analysed as well in mathematics and numerics as also in nearly any scientific discipline. Nevertheless most relevant models of nature are nonlinear, so that linear theory seems to be not applicable or in the best case only by local approximations. Here an even not new theory of functional analysis comes into play. The nonlinear operator induces in a natural way a linear one acting on the continuous functions defined on the space, where if nonlinear operator is defined. This linear Koopman operator has well known attributes as spectrum, eigenvalues, stable eigenspaces. What this means for a specific application is task for the different communities. The approach can be handled also in a numerical way, important for simulations, which has discussed already in in the Dynamic Mode Decomposition theory of Peter Schmid [12]. This is related to the Koopman operator theory, as pointed out by Igor Mesić and coworkers in [3] and Clarence Rowley and his coworkers in [4].

In that way the Ergodic Theory investigated by Ludwig Boltzmann, John von Neumann, George David Birkhoff, Bernard Osgood Koopman, Norbert Wiener,

U. Küster (✉) • R. Schneider • A. Ruopp
High Performance Computing Center Stuttgart (HLRS), Nobelstraße 19, 70569 Stuttgart, Germany
e-mail: kuester@hlrs.de


M.M. Resch et al. (eds.), *Sustained Simulation Performance 2017*,
DOI 10.1007/978-3-319-66896-3_3

Aurel Friedrich Wintner and comprehensively described in the monograph [5] for new developments comes to modern numerical applications. We try here to make some steps further to general applicability and for understanding what that implies for the analysis of the solutions of nonlinear partial differential equations. in addition, we give also the link to spectral theory of Fourier analysis.

## 2 The Koopman Operator

This is a short description of the preliminaries given in [5] and recapitulating, what has been described in [8]. Let

$$\varphi : K \longrightarrow K \tag{1}$$

be a continuous nonlinear operator on the compact space $K$ and assume $\mathscr{F} \subset C(K)$ being a linear subspace of "observables" in the continuous functions on $K$. $\mathscr{F}$ shall have the stability property

$$f \in \mathscr{F} \Rightarrow f \circ \varphi \in \mathscr{F} \tag{2}$$

that means, that an observable coupled with the operator is again in the observable space. This condition forces $\mathscr{F}$ typically to be large. Observables might be any useful functional on the space of interest as the mean pressure of a (restricted) fluid domain $\Omega$ or the evaluation operators $\delta_x$ at all points $x \in \Omega$. It might be also economic parameters describing the behaviour of models of national and global economies or of models of social science. The nonlinear operator $\varphi$ has no further restrictions. It might describe non wellposed unsteady problems, the case where trajectories are not convergent (also strange attractors), chaotic or turbulent behaviour, mixing fluids, particle systems or ensembles of trajectories for weather forecast. The operator could also be defined by an agent based system for the simulation of traffic, epidemics, social dependencies, where the agents determine their next status by the current status of some other neighbouring agents. In this case $K$ is the set product of the status of all agents with some topology and definitely not a subset of a vector space in contrast to $\mathscr{F}$. The operator $\varphi$ might even not be known explicitly, but its effect on the observables measured at a number of time steps with constant difference is present. All models are described where an operator is changing the values of the observables to a new state, as long as the iterations are not leaving the limited region of interest.

An important numerical example is the discretization of the Navier-Stokes equations on a finite set of grid points in a domain and time steps. It is even possible to understand $K$ here as the product of the status of all variables on the discretization grid together with varying boundary conditions and geometrical parameters.

By a simple mechanism the **nonlinear** operator $\varphi$ acting on a set without linear structure induces a **linear** operator on the space of observables $\mathscr{F}$. The operator $T_\varphi$ on the observables defined by

$$T_\varphi : \mathscr{F} \longrightarrow \mathscr{F} \tag{3}$$

$$f \mapsto T_\varphi f = f \circ \varphi \tag{4}$$

is named the Koopman operator of $\varphi$ on $\mathscr{F}$ [6]. Hence $T_\varphi$ is linear and continuous.

As an infinite dimensional operator $T_\varphi$ may have a (complicated) spectrum with discrete and continuous parts. We are mainly interested in the point spectrum with eigenvalues providing eigenfunctions which are elements of $\mathscr{F}$ describing behaviour in $K$. The eigenvectors or eigenfunctions $f$ are elements of the space of observables $\mathscr{F}$, not of the state space $K$ as we know it form the linear case. They fulfill Schröders functional equation [13]

$$f(\varphi q) = \lambda f(q) \quad \forall\, q \in K \tag{5}$$

The compactness of the space $K$ is forcing $|\lambda| \leq 1$ for any eigenvalue $\lambda$. For any application the meaning of these stable observables must be discussed. This might be a problem by itself.

It is a priori not clear, that eigenfunctions exist. The approach is addressing approximative eigenfunctions.

## 3 Trajectories and Observables

We study only single trajectories, even where ergodic theory [5] would allow for very general settings. But our target is to establish numerical procedures reflecting the implications of the theory at a level enabling computation. Even the trajectory might be large and dense in the space $K$. For numerical handling we assume that only the trajectory is given. We are not requesting the explicit knowledge of the operator. Also the space of observables is reduced as much as possible. We assume $h$ to be an observable or a finite dimensional vector of observables. In the latter case we assume a dotproduct $< \cdot, \cdot >$. It might be also a function of a function space. We still avoid this setting because of the difficult questions involved. But $h$ could be a function in a discrete finite space, as we have this in numerical approximations of function spaces.

Let $q_0 \in K$ be the starting point of the iteration or trajectory

$$\mathbb{N}_0 \ni k \mapsto q_k = \varphi^k q_0 \in K \tag{6}$$

Define $g_k$ as the sequence

$$g_k = h(q_k) \quad \forall\, k \in \mathbb{N}_0 \tag{7}$$

These $g_k$ are given by measurements or resulting iterations from a simulation. Because $K$ is compact and $h$ is continuous there norm has a common bound $K_h$ with $\|g_k\| \le K_h \quad \forall\, k \in \mathbb{N}_0$. They determine a matrix $G$ by ($n$ finite or infinite)

$$G = \left[g_0\; g_1 \cdots g_n\right] \tag{8}$$

The matrix $H = G^T G$ is symmetric positive semidefinite.

$$H_{j_1, j_2} = \left\langle g_{j_1}, g_{j_2} \right\rangle \quad \forall\, j_1, j_2 = 0, \cdots, n \tag{9}$$

Defining the sequence of vectors $\left(h_j\right)_j$ by overlapping the trajectory

$$h_j = \left(h\left(\varphi^j q_k\right)\right)_{k\in\mathbb{N}_0} = \left(h\left(\varphi^{j+k} q_0\right)\right)_{k\in\mathbb{N}_0} = \left(g_{j+k}\right)_{k\in\mathbb{N}_0} \tag{10}$$

we can define the space of observables $\mathscr{F}$ by

$$\mathscr{F} = \overline{LH}\left\{h_j | j \in \mathbb{N}_0\right\} \tag{11}$$

The space $\mathscr{F}$ has the stability property, necessary to define the Koopman operator on $\mathscr{F}$

$$g \in \mathscr{F} \Rightarrow g \circ \varphi \in \mathscr{F} \tag{12}$$

We define $p$ arbitrary, but fixed. Let $a = \sum_{j_1=0}^{p} \alpha_{j_1} h_{j_1} \in \mathscr{F}$ and $b = \sum_{j_2=0}^{p} \beta_{j_2} h_{j_2} \in \mathscr{F}$. The Koopman operator $T_\varphi$ acts on the space $\mathscr{F}$ via shifting on $h_j, j \in \mathbb{N}_0$.

$$T_\varphi\left(a\right) = \sum_{j=0}^{p} \alpha_j T_\varphi\left(h_j\right) = \sum_{j=0}^{p} \alpha_j h_{j+1} \tag{13}$$

or for the coefficient vector $\alpha$

$$\left[\alpha_0\; \alpha_1\; \ldots\; \alpha_p\right] \mapsto \left[0\; \alpha_0\; \alpha_1\; \ldots\; \alpha_p\right] \tag{14}$$

## 4 The Relation to Time Series Analysis

For $m \in \mathbb{N}$ the semi-sesquilinear form can be defined by

$$\ll a, b \gg_m = \ll \sum_{j_1=0}^{p} \alpha_{j_1} h_{j_1}, \sum_{j_2=0}^{p} \beta_{j_2} h_{j_2} \gg_m = \sum_{j_1, j_2=0}^{p,p} \alpha_{j_1} \overline{\beta_{j_2}} \ll h_{j_1}, h_{j_2} \gg_m \tag{15}$$

using

$$\ll h_{j_1}, h_{j_2} \gg_m := \frac{1}{m} \sum_{k=0}^{m-1} \left\langle h\left(\varphi^{j_1+k} q_0\right), h\left(\varphi^{j_2+k} q_0\right)\right\rangle$$
$$= \frac{1}{m} \sum_{k=0}^{m-1} \left\langle g_{j_1+k}, g_{j_2+k}\right\rangle \quad \forall\, 0 \le j_1 \le p, 0 \le j_2 \le p \qquad (16)$$

For all $j_1, j_2$ we have $| \ll h_{j_1}, h_{j_2} \gg_m | \le K_h^2$. The resulting matrix $\left(\ll h_{j_1}, h_{j_2} \gg_m\right)_{j_1, j_2}$ is the arithmetic mean $H^{m-1}$ of the first $m$ shifted submatrices $\frac{1}{m} \sum_{j=0}^{m-1} H_j$ of the matrix $H = G^T G$

$$H_j = H_{j:j+p, j:j+p} \qquad (17)$$

with the size $[0:p] \times [0:p]$. We have

$$\ll a, b \gg_m = < H^{m-1} \alpha, \beta > \qquad (18)$$

We assume, that the limit $\lim_{m \to \infty} \ll h_{j_1}, h_{j_2} \gg_m$ exist for all $j_1, j_2 \ge 0$.
This preassumption is not clear in case of applications.
Because for $j_1 \ge j_2$

$$H_{j_1, j_2}^{m-1} = \frac{1}{m} \sum_{k=0}^{m-1} \left\langle g_{j_1+k}, g_{j_2+k}\right\rangle = \frac{1}{m} \sum_{k=j_2}^{m-1+j_2} \left\langle g_{j_1-j_2+k}, g_k\right\rangle$$
$$= \frac{1}{m} \sum_{k=\max(0,-j_1+j_2)}^{m-1+\min(0,-j_1+j_2)} \left\langle g_{j_1-j_2+k}, g_k\right\rangle \text{ (Toeplitz matrix)} \qquad (19)$$
$$- \frac{1}{m} \sum_{k=\max(0,-j_1+j_2)}^{j_2-1} \left\langle g_{j_1-j_2+k}, g_k\right\rangle$$
$$+ \frac{1}{m} \sum_{k=m+\min(0,-j_1+j_2)}^{m-1+j_2} \left\langle g_{j_1-j_2+k}, g_k\right\rangle \qquad (20)$$

we get a decomposition in a Toeplitz matrix (the elements depend only on the difference $j_1 - j_2$) and initial and final matrices (https://en.wikipedia.org/wiki/Toeplitz_matrix), which are converging to zero as $m \to \infty$ for fixed $j_1, j_2$. For $m \gg j_1, j_2$ we end up with the relation

$$\gamma(j_1 - j_2) = \lim_{m\to\infty} \frac{1}{m} \sum_{k=0}^{m-1} \left\langle h\left(\varphi^{j_1-j_2+k} q_0\right), h\left(\varphi^k q_0\right)\right\rangle$$

$$= \lim_{m\to\infty} \frac{1}{m} \sum_{k=0}^{m-1} \left\langle h\left(\varphi^{j_1+k} q_0\right), h\left(\varphi^{j_2+k} q_0\right)\right\rangle \tag{21}$$

$$= \lim_{m\to\infty} \frac{1}{m} \sum_{k=0}^{m-1} \left\langle g_{j_1-j_2+k}, g_k\right\rangle = \lim_{m\to\infty} \frac{1}{m} \sum_{k=0}^{m-1} \left\langle g_{j_1+k}, g_{j_2+k}\right\rangle = H^{\infty}_{j_1,j_2} \tag{22}$$

for the symmetric positive semidefinite Toeplitz matrix $H^{\infty}$ related to the autocorrelation coefficient $\gamma(j_1 - j_2)$, with a modulus bounded by $K_h^2$ (https://en.wikipedia.org/wiki/Autocorrelation). For such $\gamma$ the theorem of Herglotz-Bochner (see [5] Theorem 18.6) assures the existence of a positive finite measure $\mu$ on the unit circle $\mathbb{T}$ depending on $h$ and $q_0$ with the property (see also [1])

$$H^{\infty}_{j_1,j_2} = \gamma(j_1 - j_2) = \hat{\mu}(j_1 - j_2) = \int_{\mathbb{T}} \lambda^{-(j_1-j_2)}\, d\mu(\lambda) \tag{23}$$

so that the function $\gamma$ on $\mathbb{Z}$ is the Fourier-transform of the measure $\mu$ identical to the entries of the Toeplitz matrix $H^{\infty}$. All of this is well known in time series analysis.

In terms of the generating elements of the observables $\mathscr{F}$ we get

$$\ll T_{\varphi}^{l} \sum_{j_1=0}^{p} \alpha_{j_1} h_{j_1}, \sum_{j_2=0}^{p} \beta_{j_2} h_{j_2} \gg_{\infty} = \int_{\mathbb{T}} \overline{\lambda}^{l} \sum_{j_1=0}^{p} \alpha_{j_1} \lambda^{-j_1} \overline{\sum_{j_2=0}^{p} \beta_{j_2} \lambda^{-j_2}}\, d\mu(\lambda) \tag{24}$$

showing how $T_{\varphi}$ is acting on $\mathscr{F}$ and with respect to $\mu$ for large $m$. The behaviour of small $m$, say the transition is not described.

Applied to coefficient vectors $\alpha, \beta$, the related elements , $b \in \mathscr{F}$ and the related polynoms $\lambda \mapsto \alpha(\lambda), \beta(\lambda)$

$$\ll a, b \gg_{\infty} = \langle H^{\infty}\alpha, \beta\rangle = \int_{\mathbb{T}} \alpha\left(\overline{\lambda}\right) \overline{\beta\left(\overline{\lambda}\right)}\, d\mu(\lambda) \tag{25}$$

Following [1] Theorem 2.1 we see that the measure $\mu$(23) is the weak limit of the sequence of measures $\mu_m$ given by the density for the Lebesgue measure $d\lambda$ on the unit circle

$$\int_{\mathbb{T}} r(\lambda)\, d\lambda = \int_0^{2\pi} r(\exp i\,\phi)\, d\phi \text{ for all integrable } r \tag{26}$$

so that

$$\mu_m(E) = \frac{1}{2\pi} \int_E d_m(\lambda)\, d\lambda \quad \forall \text{ measurable sets } E \subset \mathbb{T} \tag{27}$$

with the density function

$$\mathbb{T} \ni \lambda \mapsto d_m(\lambda) = \frac{1}{m} \left\| \sum_{k=0}^{m-1} g_k \, \overline{\lambda}^k \right\|^2 \tag{28}$$

which approximates the spectral density of the eigenvalues of modulus 1 and can be calculated on a computer. The function is not bounded with respect to $m$ even for constant $g_k = g_0$.

Furthermore we have the relation (using, that $\int_{\mathbb{T}} \lambda^l \, d\lambda = 0$ for $l \neq 0$ and $\int_{\mathbb{T}} \lambda^l \, d\lambda = 2\pi$ for $l = 0$).

$$\frac{1}{2\pi} \int_{\mathbb{T}} \lambda^{-(j_1 - j_2)} \, d_m(\lambda) \, d\lambda = \frac{1}{2\pi} \frac{1}{m} \sum_{k_1, k_2 = 0}^{p} \langle g_{k_1}, g_{k_2} \rangle \int_{\mathbb{T}} \lambda^{-(j_1 - j_2) + k_1 - k_2} \, d\lambda \tag{29}$$

$$= \frac{1}{m} \sum_{k = \max(0, -j_1 + j_2)}^{m - 1 - \min(0, -j_1 + j_2)} \left\langle g_{j_1 - j_2 + k}, g_k \right\rangle \text{ for } j_1 \geq j_2 \tag{30}$$

$$= \frac{1}{m} \sum_{k = \max(0, j_1 - j_2)}^{m - 1 + \min(0, j_1 - j_2)} \left\langle g_k, g_{j_2 - j_1 + k} \right\rangle \text{ for } j_2 \geq j_1 \tag{31}$$

The last two identities reproduce the Toeplitz matrix appearing in Eq. (19). The elements are bounded by

$$\left| \frac{1}{2\pi} \int_{\mathbb{T}} \lambda^{-(j_1 - j_2)} \, d_m(\lambda) \, d\lambda \right| \leq K_h^2 \quad \forall \, 0 \leq j_1, j_2 \leq p \tag{32}$$

The density function (28) divided by $m$ is the square of the norm of the discrete Fourier backtransform of the finite sequence $\left[ g_0 \; g_1 \; \cdots \; g_{m-1} \right]$ which again is bounded by $K_h^2$. Additionally we recognize that

$$< H^{m-1}\alpha, \beta > - \frac{1}{2\pi} \int_{\mathbb{T}} \sum_{j_1 = 0}^{p} \alpha_{j_1} \overline{\lambda^{j_1}} \sum_{j_2 = 0}^{p} \overline{\beta_{j_2}} \lambda^{j_2} \, d_m(\lambda) \, d\lambda \overset{m \to \infty}{\longrightarrow} 0 \tag{33}$$

**Definition 1** For a polynom $c$ define the norm $\|c\|_{\infty, \mathbb{T}} = \max_{\lambda \in \mathbb{T}} |c(\lambda)|$ . This is the $H^\infty$ Hardy norm of the polynom $\|c\|_{H^\infty}$ on the unit circle (https://en.wikipedia.org/wiki/Hardy_space).

**Proposition 1** *Assume, the finite measure* $\mu$ *(23) is discrete and that* $\epsilon > 0$ *is given.*

1. *$\mu$ must be the sum of at most countable number of point measures*

$$\mu = \sum_{l=1}^{\infty} \rho_l \, \delta_{\{\lambda_l\}} \text{ with } |\lambda_l| = 1 \quad \forall \, l \tag{34}$$

*with decreasing weights $\rho_l > 0$.*

*It exists $l_\epsilon$, so that the last part of the sum is small $\sum_{l=l_\epsilon+1}^{\infty} \rho_l < \epsilon$.*

2. *Let $c = c_\epsilon$ be a polynom coefficient vector $c$, so that $c\left(\overline{\lambda_l}\right) = 0$ for $l = 1, \cdots, l_\epsilon$. Then*

$$\frac{\langle H^\infty c, c\rangle}{\|c\|^2_{\infty,\mathbb{T}}} = \frac{1}{\|c\|^2_{\infty,\mathbb{T}}} \int_{\mathbb{T}} \left| \sum_{j=0}^{p} c_j \lambda^{-j} \right|^2 d\mu\left(\lambda\right) = \frac{1}{\|c\|^2_{\infty,\mathbb{T}}} \sum_{l=1}^{\infty} \rho_l \left|c\left(\overline{\lambda_l}\right)\right|^2$$

$$= \frac{1}{\|c\|^2_{\infty,\mathbb{T}}} \sum_{l=l_\epsilon+1}^{\infty} \rho_l \left|c\left(\overline{\lambda_l}\right)\right|^2 \leq \epsilon \tag{35}$$

3. *By definition of $H^\infty$ there exists an $m_0$ so that*

$$\frac{\frac{1}{m}\left\langle \sum_{j=0}^{m-1} H_j c, c\right\rangle}{\|c\|^2_{\infty,\mathbb{T}}} \leq 2\epsilon \quad \forall \, m \geq m_0 \tag{36}$$

4. *Multiplying $c$ by any other non zero polynom $b$ would maintain the roots of $c$ and therefore also this estimate for $c * b$.*
5. *Because the $l_2$-norm of the polynom coefficient vector $c$ is identical to the $H^2$ Hardy space norm of the polynom $c$ which itself is lower or equal to the $H_\infty$ Hardy space norm of the polynom*

$$\|c\|_2^2 = \|c\|^2_{H_2} = \frac{1}{2\pi} \int_{\mathbb{T}} \left| \sum_k c_k \lambda^k \right|^2 d\lambda \leq \frac{1}{2\pi} \int_{\mathbb{T}} \max_{\lambda' \in \mathbb{T}} \left|c\left(\lambda'\right)\right|^2 d\lambda = \|c\|^2_{\infty,\mathbb{T}} \leq \|c\|_1^2 \tag{37}$$

*the Rayleigh quotient of $c$ is an upper estimate of (35)*

$$\frac{\langle H^\infty c, c\rangle}{\|c\|^2_{\infty,\mathbb{T}}} \leq \frac{\langle H^\infty c, c\rangle}{\|c\|_2^2} \tag{38}$$

*meaning, that for a polynom coefficient vector $c$ with $\frac{\langle H^\infty c, c\rangle}{\|c\|_2^2} \leq \epsilon$, we have also an estimate for (35).*

It is not clear , that $\|c_\epsilon\|_{\infty,\mathbb{T}}$ remains limited for $\epsilon \to 0$. In numerical tests we got the impression, that this is the case if the roots are nearly uniformly distributed near the unit circle.

For a measure $\mu$ (23) with continuous parts, the existence of a polynomial approximation of nullfunctions of $H^\infty$ is not yet clear.

The analysis as far shows the relation to a Fourier analysis by the Herglotz-Bochner theorem. That means it shows the behaviour of a long running sequence neglecting the damped parts of the sequence. Only the spectral part on the unit circle $\mathbb{T}$ is relevant in this context and not the inner part of the unit circle describing the part, which will vanish during the iteration. Nevertheless a numerical scheme must show the correct results in the terms described in the previous section.

The following examples will give insight into typical situations for given sequences $(g_k)_{k=0,1,2,\ldots}$. All these can be part of a single sequence.

*Example 1 (Converging Sequences)* If the iteration $h\left(\varphi^k\right) \overset{k\to\infty}{\to} h\left(q_*\right)$ converges, the measure $\mu$ in (23) is the single point measure at 1.

To see this

$$\begin{aligned}
&\left|\frac{1}{m}\sum_{k=k_0}^{m-1+k_0}\left\langle h\left(\varphi^{j_1+k}q_0\right), h\left(\varphi^{j_2+k}q_0\right)\right\rangle - \left\langle h\left(q_*\right), h\left(q_*\right)\right\rangle\right| \\
&\le \left|\frac{1}{m}\sum_{k=k_0}^{m-1+k_0}\left\langle h\left(\varphi^{j_1+k}q_0\right) - h\left(q_*\right), h\left(\varphi^{j_2+k}q_0\right) - h\left(q_*\right)\right\rangle\right| \\
&+ \left|\frac{1}{m}\sum_{k=k_0}^{m-1+k_0}\left\langle h\left(\varphi^{j_1+k}q_0\right) - h\left(q_*\right), h\left(q_*\right)\right\rangle\right| \\
&+ \left|\frac{1}{m}\sum_{k=k_0}^{m-1+k_0}\left\langle h\left(q_*\right), h\left(\varphi^{j_2+k}q_0\right) - h\left(q_*\right)\right\rangle\right|
\end{aligned} \tag{39}$$

so that for given $\epsilon > 0$ and appropriate $k_0$ , that $\|h\left(\varphi^{j+k}q_0\right) - h\left(q_*\right)\| < \epsilon$ for all $k, j \ge 0$ with $k \ge k_0$ we find

$$\left|\lim_{m\to\infty}\frac{1}{m}\sum_{k=k_0}^{m-1+k_0}\left\langle h\left(\varphi^{j_1+k}q_0\right), h\left(\varphi^{j_2+k}q_0\right)\right\rangle - \left\langle h\left(q_*\right), h\left(q_*\right)\right\rangle\right| \le \epsilon^2 + \epsilon\|h\left(q_*\right)\| \tag{40}$$

and therefore $\lim_{m\to\infty}\frac{1}{m}\sum_{k=0}^{m-1}\left\langle h\left(\varphi^{j_1+k}q_0\right), h\left(\varphi^{j_2+k}q_0\right)\right\rangle = \lim_{m\to\infty}\frac{1}{m}\sum_{k=k_0}^{m-1+k_0}\left\langle h\left(\varphi^{j_1+k}q_0\right), h\left(\varphi^{j_2+k}q_0\right)\right\rangle = \|h\left(q_*\right)\|^2$. That means

$$\|h\left(q_*\right)\|^2 = \int_{\mathbb{T}} \lambda^{-(j_1-j_2)}\, d\mu\left(\lambda\right) \quad \forall\, j_1, j_2 \ge 0 \tag{41}$$

which is possible only for the point measure $\mu = \|h\left(q_*\right)\|^2\,\delta_{\{1\}}$.

This example shows the measure $\mu$ (23) for a practically relevant case, but which is of less interest in our context.

*Example 2 (Besicovitch Sequences)* Assume, that the values $g_k = \sum_{l=1}^{p} v_l \lambda_l^k$ $\forall\ k \in \mathbb{N}_0$ are given by decomposition in modes with all $\lambda_l$ pairwise different and $|\lambda_l| \leq 1$. If for all $|\lambda_l| = 1$, the decomposition is a so called Besicovitch sequence [2]. Then

$$H_{j_1,j_2}^{m-1} = \frac{1}{m} \sum_{k=0}^{m-1} \langle g_{j_1+k}, g_{j_2+k} \rangle = \frac{1}{m} \sum_{k=0}^{m-1} \sum_{l_1,l_2} \langle v_{l_1}, v_{l_2} \rangle\ \lambda_{l_1}^{j_1+k} \overline{\lambda_{l_2}}^{j_2+k}$$

$$= \sum_{l_1,l_2} \langle v_{l_1}, v_{l_2} \rangle\ \lambda_{l_1}^{j_1} \overline{\lambda_{l_2}}^{j_2}\ \frac{1}{m} \sum_{k=0}^{m-1} \lambda_{l_1}^{k} \overline{\lambda_{l_2}}^{k} \tag{42}$$

$$= \sum_{l_1,l_2} \langle v_{l_1}, v_{l_2} \rangle\ \lambda_{l_1}^{j_1} \overline{\lambda_{l_2}}^{j_2}\ \frac{1}{m}\ \frac{1 - \left(\lambda_{l_1} \overline{\lambda_{l_2}}\right)^m}{1 - \lambda_{l_1} \overline{\lambda_{l_2}}} \tag{43}$$

The last term converges to zero for $m \to \infty$ if $\lambda_{l_1} \overline{\lambda_{l_2}} \neq 1$ because $|\lambda_{l_1} \overline{\lambda_{l_2}}| \leq 1$. The convergence is relatively slow. For $\lambda_{l_1} \overline{\lambda_{l_2}} = 1$ we simply have $\frac{1}{m} \sum_{k=0}^{m-1} \lambda_{l_1}^k \overline{\lambda_{l_2}}^k = 1$. Therefore

$$H_{j_1,j_2}^{\infty} = \lim_{m \to \infty} \frac{1}{m} \sum_{k=0}^{m-1} \langle g_{j_1+k}, g_{j_2+k} \rangle = \sum_{l} \langle v_l, v_l \rangle\ \lambda_l^{j_1-j_2} \tag{44}$$

Thus the measure $\mu$ (23) is discrete and given by

$$\mu = \sum_{l} \|v_l\|^2\ \delta_{\{\overline{\lambda_l}\}} \tag{45}$$

meaning, that the values $\overline{\lambda_l}$ of the Besicovitch sequence determine the measure $\mu$. Because the measure $\mu$ is finite, the sum $\sum_l \|v_l\|^2$ must be bounded. The assumption $|\lambda_l| \leq 1$ is essential for the construction. Modes with $|\lambda_l| < 1$ are disappearing for $k \to \infty$. The measure $\mu$ (23) cannot represent these.

Let $c$ be an polynom coefficient vector and define the linear combination $\sum_{k=0}^{p} c_k g_{k+j} = \sum_l v_l \lambda_l^j c(\lambda_l)$. We assume, that a finite number of $\lambda_l$ are roots of the polynom $\lambda \mapsto c(\lambda)$. Estimating the linear combination by

$$\| \sum_{k=0}^{p} c_k g_{k+j} \| \leq \sum_{l} \|v_l\| |\lambda_l^j| |c(\lambda_l)| \tag{46}$$

$$\leq \sum_{l, c(\lambda_l) \neq 0} \|v_l\| |c(\lambda_l)| \tag{47}$$

$$\leq \sum_{l, c(\lambda_l) \neq 0} \|v_l\| \max_{|\lambda|=1} |c(\lambda)| \tag{48}$$

so that for finite $\sum_l \|v_l\| < \infty$ it is possible to define the roots of $c$ in way, that the remaining part is arbitrary small independent on the index $j$ as long as $\max_{|\lambda|=1} |c(\lambda)|$ remains bounded.

*Example 3 (Impact of Continuous Segments)* Assume, that the values $g_k$ for $k \in \mathbb{N}_0$ have a linear decomposition containing also continuous parts of the following kind as summands.

$$g_k = \sum_m \int_{[\rho_m,\sigma_m]} \lambda^k f_m(\lambda)\ d\lambda \quad \forall\, k \in \mathbb{N}_0 \tag{49}$$

For simplicity we consider one element of the sum.

The function $\lambda \mapsto f(\lambda)$ should be continuously differentiable. $[\rho, \sigma]$ is a segment on the unit circle. Because of the definition of (26) (remark the difference to the usual product rule)

$$\int_{[\rho,\sigma]} \lambda^k f(\lambda)\ d\lambda = \left[\frac{1}{i\,k}\,\lambda^k f(\lambda)\right]_\rho^\sigma - \int_{[\rho,\sigma]} \frac{1}{k}\,\lambda^{k+1}\ \partial_\lambda f(\lambda)\ d\lambda \tag{50}$$

these terms converge to 0 for $k \to \infty$ with $\frac{1}{k}$ and not exponentially as in the case of a discrete atomic measure for a point $\lambda$ with $|\lambda| < 1$ . Even a constant $f$ will be interesting.

By these contiguous segments continuous parts of the spectrum can be formulated for numerical purposes. Used as density for the Lebesgue measure on the unit circle they are special examples for so called Rajchman measures, which $n$th Fourier Transform vanish for $n \to \infty$ (https://en.wikipedia.org/wiki/Rajchman_measure).

## 5 Approximation of an $\lambda$-Eigenmode Along a Trajectory

As before we assume here $\varphi$ as an operator and a sequence $f = (f_j)_{j=0,1,\cdots}$ with $f_j = f(\varphi^j q_0)$ of scalars or vectors as in (10) or even functions as elements of a function space $C(\Omega, \mathbb{R}^s)$, $f_j$ are the elements of an iteration observed by a vector of observables $f$. $s$ might be any natural number given by a number of components in the analysed process defined by the operator $\varphi$. $s = 4$ in case of the incompressible Navier-Stokes equations with three velocity components and the pressure. To simplify the problem we handle only discretized versions with a finite set of discretization points $\Omega$ which makes the function space finite dimensional. $j$ is the time step number of the time discretization given by the operator $\varphi$ in this case. We define finite complex linear combinations of values along the trajectory starting at each step $j$. The coefficients are given by the vector $\alpha = (\alpha_k)_{k=0,\cdots,p-1}$ and do not dependent on $j$. The coefficients are understood also as coefficients of a

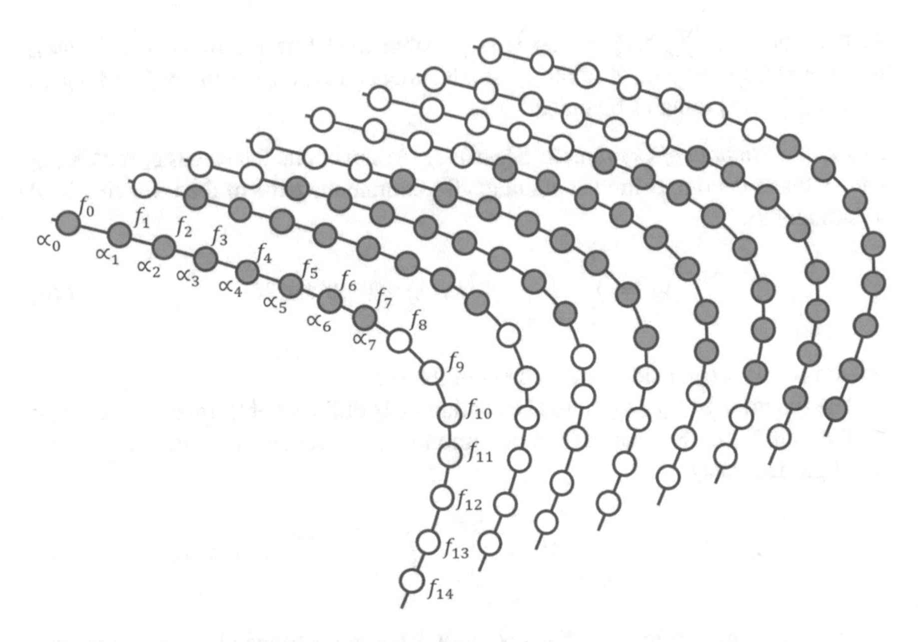


**Fig. 1** Moving linear combination of values along a trajectory

polynom $\lambda \mapsto \alpha(\mu) = \sum_{k=0}^{p-1} \alpha_k \mu^k$. We fix here a value $\lambda$ which must not be a root of this polynom and define the sequence

$$\widehat{f}_j^{\alpha,\lambda} = \frac{\sum_{k=0}^{p-1} \alpha_k f_{j+k}}{\sum_{k=0}^{p-1} \alpha_k \lambda^k} \quad \forall j = 0, 1, 2, \cdots \tag{51}$$

Figure 1 shows the linear combinations on the trajectory for $p-1 = 8$ elements moving along 7 starting points $j$. This approach is motivated by the fact, that for a sequence $\left(f_j\right)_j$, already fulfilling the rule $f_j = \lambda^j f_0$, we conserve this property $\widehat{f}_j^{\alpha,\lambda} = \lambda^j f_0$. That means we conserve Schröders equation $f(\varphi q) = \lambda f(q)\ \forall q$, or in other words have a Koopman eigenvector.

The error $\epsilon^{\alpha,\lambda}$ of the pair $\left(\lambda, \widehat{f}^{\alpha,\lambda}\right)$ of being a Koopman eigenvalue-eigenvector pair is given by

$$\epsilon_j^{\alpha,\lambda} = -\lambda \widehat{f}_j^{\alpha,\lambda} + \widehat{f}_{j+1}^{\alpha,\lambda} = \frac{1}{\alpha(\lambda)} \left( -\sum_{k=0}^{p-1} f_{j+k}\, \lambda \alpha_k + \sum_{k=0}^{p-1} f_{j+1+k}\, \alpha_k \right) \tag{52}$$

$$= \frac{1}{\alpha(\lambda)} \left( -f_j\, \lambda \alpha_0 + \sum_{k=1}^{p-1} f_{j+k} \left(-\lambda \alpha_k + \alpha_{k-1}\right) + f_{j+p}\, \alpha_{p-1} \right) \tag{53}$$

$$= \frac{1}{\alpha(\lambda)} \sum_{k=0}^{p} f_{j+k}\, c_k \quad \forall j = 0, 1, \cdots \tag{54}$$

with the polynom coefficient vector

$$\begin{aligned} c_0 &= -\lambda\alpha_0 \\ c_k &= -\lambda\alpha_k + \alpha_{k-1} \quad \forall\, k = 1, \cdots, p-1 \\ c_p &= \qquad\quad \alpha_{p-1} \end{aligned} \tag{55}$$

*Example 4 (Wiener-Wintner)* An example are the sums in the Wiener-Wintner theorem [5]

$$\tilde{f}_\omega\left(q_j\right) \approx \frac{1}{p}\sum_{k=0}^{p-1} f\left(\varphi^k q_j\right) e^{i2\pi\,\omega k} \quad \forall\, j \in \mathbb{N}_0 \tag{56}$$

where $\alpha_k = e^{i2\pi\,\omega k}$ for all $k$, the eigenvalue $\lambda = e^{-i2\pi\,\omega}$ and $\alpha(\lambda) = p$. For $p \to \infty$ they approximate the eigenmode $\tilde{f}_\omega$ for $\lambda$.

$c$ is given by $c = \left[-e^{-i2\pi\,\omega}\; 0 \cdots 0\; e^{i2\pi\,\omega(p-1)}\right]$ and the approximation error by $\epsilon_j^{\alpha,\lambda} = \frac{1}{p}\left(-f_j e^{-i2\pi\,\omega} + f_{j+p} e^{i2\pi\,\omega(p-1)}\right)$ which is small for bounded $f$, if $p$ is large.

As a polynom, $c$ is the product of polynom $\alpha$ and the linear divisor given by $\mu \mapsto \mu - \lambda$. Understood as polynom coefficient vector $c$ is the convolution $c = \alpha * \begin{bmatrix} -\lambda \\ 1 \end{bmatrix}$.

By definition, $\lambda$ is a root of the polynom $\mu \mapsto c(\mu)$.

For a matrix $\mathfrak{A}_p(c)$ build by the repeatedly shifted vector $c$ we request equivalently

$$\epsilon^{\alpha,\lambda} = \frac{1}{\alpha(\lambda)}\left[f_0, f_1, f_2, \cdots\right] \begin{bmatrix} c_0 & & & \\ c_1 & c_0 & & \\ c_2 & c_1 & c_0 & \\ \cdot & c_2 & c_1 & \cdot \\ \cdot & \cdot & \cdot & \cdot \\ c_p & c_{p-1} & c_{p-2} & \cdot \\ & c_p & c_{p-1} & \cdot \\ & & c_p & \cdot \\ & & & \cdot \end{bmatrix} = \frac{1}{\alpha(\lambda)} f\, \mathfrak{A}_p(c) \overset{!}{\approx} 0 \tag{57}$$

The existence of a polynom coefficient vector $c$ with this property is a necessary condition for the existence of approximative eigenvectors. Using the same notation we have $\widehat{f}^{\alpha,\lambda} = \frac{1}{\alpha(\lambda)} f\, \mathfrak{A}_{p-1}(\alpha)$ for the approximative eigenmode (51) .

The sequence $\widehat{f}^{\alpha,\lambda}$ will be an approximative eigenmode of the underlying iteration operator if and only if $\|\widehat{f}^{\alpha,\lambda}\| \gg \|\epsilon^{\alpha,\lambda}\|$ and $\|\epsilon^{\alpha,\lambda}\| \approx 0$. In the following we assume that $\epsilon^{\alpha,\lambda}$ is small.

If a vector $c$ with the property (57) has been found, it provides $\lambda$ as root of the polynom $c$. Furthermore, all roots $\lambda_l$ of $c$ together with the related polynom coefficient vectors $\alpha_l$ for $l = 1, \cdots, p$ are candidates for approximating Koopman modes, as long they are not multiple roots and as long $|\lambda_l| \leq 1$ . These pairs all share the same approximation quality $\frac{1}{\alpha(\lambda_l)}f\,\mathfrak{A}_p(c)$ depending solely on the fraction $\frac{1}{\alpha(\lambda_l)}$. By providing a single coefficient vector $\alpha$, we get many other coefficient vectors $\alpha_l$ by dividing the polynom $c$ by the linear divisors of the different roots. If $|\alpha_l\,(\lambda_l)\,|$ is large, we expect a good approximation quality. The degree $p$ has to be as small as possible.

## 5.1 Determining the Polynom Coefficient Vector c

Given are a finite sequence of vectors $G = \left[g_0\; g_1\; \dots\; g_n\right]$ . To find a polynom coefficient vector $c$ with small approximation error

$$\epsilon^{\alpha,\lambda} = G\,\mathfrak{A}_p(c) \approx 0,$$

we apply the following procedure. For any given $c$ determine the minimal number $\rho_c \geq 0$ so that

$$\mathfrak{A}(c)^* H\,\mathfrak{A}(c) \leq \rho_c\;\mathfrak{A}(c)^*\mathfrak{A}(c) \tag{58}$$

$\rho_c$ is the largest eigenvalue of the generalized eigenvalue problem. Determine $c$ with fixed degree $p$ so that $\rho = \rho_c$ is minimal and that the roots of the polynom defined by $c$ have modulus not more than 1. This can be reached by an iterative process following the theorem of Rellich [10], that for a real symmetric parametrized matrix eigenvalues and eigenvectors depend analytically on the parameter. During the iteration the roots of $c$ are tested and changed, if there modulus exceeds 1.

The value of $\rho$ determines the error of the approximation.

Projecting to the $j$th row and column of both sides of the matrix inequality we find

$$\left\langle H_j\, c, c\right\rangle \leq \rho\;\|c\|^2 \tag{59}$$

for the $j$th submatrix $H_j$ (17).

Taking the mean over the $m = n - p + 1$ first submatrices $H_j$ we find for $H^{m-1}$ as in (18)

$$\left\langle H^{m-1} c, c\right\rangle \leq \rho\|c\|^2 \tag{60}$$

So we get by (38) and (35) the estimate for $m$ which is sufficiently large

$$\frac{\left\langle H^{\infty} c, c\right\rangle}{\|c\|^2_{\infty,\mathbb{T}}} \approx \frac{\left\langle H^{m-1} c, c\right\rangle}{\|c\|^2_{\infty,\mathbb{T}}} \leq \frac{\left\langle H^{m-1} c, c\right\rangle}{\|c\|^2_2} \leq \rho \tag{61}$$

Minimizing $\frac{\left\langle H^{m-1} c,c\right\rangle}{\|c\|^2_{\infty,\mathbb{T}}}$ directly with respect to $c$ might result in a better $c$.

There are other alternatives to calculate the vector $c$. The whole procedure is related to DMD of Peter Schmid [12].

## 5.2 Roots and Pseudo-Eigenvectors

The roots $\lambda_l$ for all $l = 1, \cdots, p$ of the polynom $c$ are determined by the solution of the eigenvalues of the $c$-companion matrix.

For each $\lambda_l$ with $|\lambda_l| \leq 1$ a "pseudo-eigenvector" $\alpha_l$ is determined by factorizing the linear divisor of root $\lambda_l$ out of c

$$\alpha_l(\lambda)\ (\lambda - \lambda_l) = c(\lambda) \quad \forall\, \lambda \in \mathbb{C} \tag{62}$$

We assume, that there are no multiple roots of $c$, so that $\alpha_l(\lambda) \neq 0$. The vectors $\alpha_l$ with degree $p-1$ define the linear combinations (51) along the trajectory delivering approximate Koopman eigenvectors for the eigenvalues $\lambda_l$.

The matrix consisting on the vectors $\frac{\alpha_l}{\alpha(\lambda_l)}$ is the inverse of the Vandermonde matrix (https://en.wikipedia.org/wiki/Vandermonde_matrix) defined by the eigenvalues $\lambda_l$.

The approximative eigenvectors are given by

$$v_l = G\,\frac{\alpha_l}{\alpha(\lambda_l)} \tag{63}$$

The vectors with complex elements are also named Koopman modes [3].

For small $\rho$ (61) we get the following approximation

$$g_k \approx \sum_{l=1}^{p} v_l\, \lambda_l^{\,k} \quad \forall\, k \tag{64}$$

which is Besicovitch like sequence including potentially some $\lambda_l$ with $|\lambda_l| < 1$.

## 5.3 Handling Intermediate Data Steps

If the number of data steps is large, or if the data from step to step are very slowly changing, or if the data size per step is large (e.g. 100,000 steps of a weather simulation), it is reasonable to analyse only intermediate steps, e.g. every $m$th out of $n$ steps. There is basically no difference to the given approach. How to analyse the influence of the intermediate data on the eigenvalues? This can be done by enlarging the matrix $G$ in the following way:

$$\widehat{G} = \begin{bmatrix} G_{0+0} & G_{m+0} & G_{2m+0} & \cdots & G_{qm+0} \\ G_{0+1} & G_{m+1} & G_{2m+1} & \cdots & G_{qm+1} \\ G_{0+2} & G_{m+2} & G_{2m+2} & \cdots & G_{qm+2} \\ \vdots & \vdots & \vdots & \cdots & \vdots \\ G_{0+m-1} & G_{m+m-1} & G_{2m+m-1} & \cdots & G_{qm+m-1} \end{bmatrix} \quad (65)$$

$q$ is defined so that $qm + m - 1 \leq n$. Every line out of $m$ consists on intermediate steps with distance $m$. The next line is shifted by 1. Be aware, that the matrix size is essentially the same as before. The degree $qm$ vector $c = \left[c_0 \, 0_{m-1} \, c_m \, 0_{m-1} \, c_{2m} \, 0_{m-1} \cdots 0_{m-1} \, c_{qm}\right]$ where $0_{m-1}$ represents a vector of $m-1$ zeros. After suppressing these zeros $\widehat{G}\, c \approx 0$ is handled as before. $c$ leads to a polynom

$$\lambda \mapsto \sum_{k=0}^{q} c_{km} \left(\lambda^m\right)^k \quad (66)$$

The $qm$ roots of this polynom are $\lambda_l^m = \lambda_{l,j}^m$ where

$$\lambda_{l,j} = \lambda_{l,0} \, \exp\left(i \, 2\pi \, \frac{j}{m}\right) \quad \forall j = 0, 1, \cdots, m-1, \text{ and } l = 1, \cdots, q \quad (67)$$

If $\lambda_{l_1,j}^m \neq \lambda_{l_2,j}^m$ for $l_1 \neq l_2$ than all the $m$ roots belonging to both values are different. These roots are related to pseudoeigenvectors $w_{l,j}$ which are calculated in the following way. Similar to $c$ let $w_l$ be the polynom coefficient vector of degree $(q-1)\, m$ $w_l = \left[w_{l,0} \, 0_{m-1} \, w_{l,m} \, 0_{m-1} \, w_{l,2m} \, 0_{m-1} \cdots w_{l,qm}\right]$ defined by the factorization $w_l = c / \left[-\lambda_l^m \, 0_{m-1} \, 1\right]$. The complete pseudoeigenvectors are then given by the convolution

$$w_{l,j} = \frac{1}{m} \left( \frac{1}{\lambda_{l,0}^k} \, \exp\left(-i \, 2\pi \, \frac{j\,k}{m}\right)\right)_{k=0,\cdots,m-1} * w_l \quad \forall j = 0, \cdots, m-1 \quad (68)$$

The eigenmodes $v_{l,j}$ are calculated by

$$u_{l,j} = \frac{1}{m} \left(\exp\left(-i \, 2\pi \, \frac{j\,k}{m}\right)\right)_{k=0,\cdots,m-1} diag\left(\frac{1}{\lambda_{l,0}^k}\right)_k \widehat{G}\, w_l / w_l\left(\lambda_l^m\right) \quad (69)$$

The first part is a discrete Fouriertransform and is identical for all different $l = 1, \cdots, q$. The second part $\widehat{G}\, w_l / w_l\left(\lambda_l^m\right)$ is constant with respect to $j$ and represents the whole space belonging to the eigenvalues given by $\lambda_l^m$. The modes $u_{l,j}$ for fixed $l$ are not related to each other. Some of them might be nearly vanishing. The calculation requires the special definition of $\widehat{G}$ with replicated shifted lines.

## 6 The $\lambda$-Eigenmode Mapping Operator

For a given pair $(\lambda, \alpha)$ we denominate the map $\widehat{\bullet}^{\lambda}$ defined by approximative eigenmode (51)

$$\widehat{\bullet}^{\lambda} : f \mapsto \widehat{f}^{\lambda} = \widehat{f}^{\alpha,\lambda} \tag{70}$$

the $\lambda$-eigenmode mapping operator. This linear operator can be applied to a sequence of scalars or vectors or functions or vector fields in the appropriate spaces.

Under reasonable conditions we found in our simulations (no proof) empirically for the pairs $(\alpha, \lambda)$, that

$$\frac{1}{|\alpha(\lambda)|} \sum_{k=0}^{p-1} |\alpha_k| = \frac{\|\alpha\|_1}{|\alpha(\lambda)|} = O(1) \tag{71}$$

This helps in analysing the behaviour in combination with diverse operators. This property together with an uniformly bounded approximation-error $\epsilon^{\alpha,\lambda}$ (57) shows the following properties.

In the case of continuous or differentiable or integrable functions $f$ the operator $\widehat{\bullet}^{\lambda}$ is linear and commutes with limits and (discrete) differentiable and integration operators, substantial for the definition of partial differential equations, e.g.

$$\begin{aligned}
\widehat{af+bg}^{\lambda} &= a\widehat{f}^{\lambda} + b\,\widehat{g}^{\lambda} \\
\lim_{n\to\infty} \widehat{f_n}^{\lambda} &= \widehat{\lim_{n\to\infty} f_n}^{\lambda} \\
\widehat{\text{grad}\, f}^{\lambda} &= \text{grad}\,\widehat{f}^{\lambda} \\
\widehat{\Delta f}^{\lambda} &= \Delta \widehat{f}^{\lambda} \\
\widehat{\text{D}\, v}^{\lambda} &= \text{D}\,\widehat{v}^{\lambda} \\
\widehat{\text{div}\, v}^{\lambda} &= \text{div}\,\widehat{v}^{\lambda} \\
\widehat{\text{rot}\, v}^{\lambda} &= \text{rot}\,\widehat{v}^{\lambda} \\
\widehat{\int_V f(x)dx}^{\lambda} &= \int_V \widehat{f}^{\lambda}(x)\,dx \\
\widehat{\oint_{\partial V} < v(x), df(x) >}^{\lambda} &= \oint_{\partial V} < \widehat{v}^{\lambda}(x)\,df(x) >
\end{aligned}$$

all these elements are approximative eigenmodes for the eigenvalue $\lambda$ as long as approximation-error $\epsilon^{\alpha,\lambda}$ (57) is small. The approximation error will surely changed by these operations. If the finite sums by coefficients would be changed to infinite sums, additional restrictions have to be expected.

Assume a time dependent solution of a partial differential equation with boundary conditions given. The $\lambda$-eigenmode mapping operator $\widehat{\bullet}^\lambda$ can be applied to the trajectory of boundary conditions as well. A timewise constant boundary condition $b$ is an eigenvector for $\lambda = 1$.

## 6.1 Incompressible Navier-Stokes Equations as Example

The Navier-Stokes equations can be defined in integral or differential form. The difference is not important here. In differentiable form

$$\operatorname{div} v = 0 \tag{72}$$

$$\partial_t\, v = -\operatorname{div} v \otimes v - \frac{1}{\rho}\operatorname{grad} p + \nu \Delta v \tag{73}$$

$\operatorname{div} v \otimes v$ is the sole nonlinear term using the local tensor product of the velocity field. The density is here simply 1. $\nu$ is the kinematic viscosity.

We discretize the time derivative in a simple way, which is here not relevant, and get for $k = 0, \cdots, p-1$ and $j = 0, 1, 2, \cdots$

$$\operatorname{div} v_{j+k} = 0 \tag{74}$$

$$v_{j+k+1} = v_{j+k} + \Delta t\left(-\operatorname{div}(v \otimes v)_{j+k} - \frac{1}{\rho}\operatorname{grad} p_{j+k} + \nu \Delta v_{j+k}\right) \tag{75}$$

The operators in 3D-space are to be understood as differentiable operators or their discretization.

Applying the $\lambda$-eigenmode mapping operator $\widehat{\bullet}^\lambda$ with respect to eigenvalue $\lambda$ to this equation system

$$\operatorname{div} \widehat{v_j}^\lambda = 0$$

$$\frac{1}{\Delta t}\left(\lambda \widehat{v_j}^\lambda - \widehat{v_j}^\lambda\right) = -\operatorname{div}\widehat{\left(v_j \otimes v_j\right)}^\lambda - \frac{1}{\rho}\operatorname{grad}\widehat{p_j}^\lambda + \nu \Delta \widehat{v_j}^\lambda \quad \forall\, j = 0, 1, 2, \cdots \tag{76}$$

After dividing by $\lambda^j$ this gives an approximate decomposition into time independent complex components.

Remark, that the equation reflects the actual spacial discretizations of grad, div, $\Delta$ as long as these are linear. E.g. the approximate eigenvector of the velocity field is divergence free.

Remark further, that

$$\widehat{(v_j \otimes v_j)}^{\lambda} \neq \left(\widehat{v_j}^{\lambda} \otimes \widehat{v_j}^{\lambda}\right) \quad \forall\, j = 0, 1, 2, \cdots \tag{77}$$

This is the nonlinear term and couples eigenmodes of different eigenvalues (but not all).

To get all of this with a small approximation-error $\epsilon^{\alpha,\lambda}$ (57) we assume that the relevant entities are small after multiplication by $\mathfrak{A}_p(c)$ from the right

$$\begin{bmatrix} v \\ v \otimes v \\ p \end{bmatrix} \mathfrak{A}_p(c) = G\, \mathfrak{A}_p(c) \overset{!}{\approx} 0 \tag{78}$$

The operator $G$ consists on the sequences of all iterations $v = \left[v_j\right]_j$, $v \otimes v = \left[v_j \otimes v_j\right]_j$ and $p = \left[p_j\right]_j$. Remark that we are using (discrete) functions. Practically it turns out, that the term $v \otimes v$ involving many variables, is not necessary. We still don't know, why and under what circumstances.

In the case of a small approximation-error $\epsilon^{\alpha,\lambda}$ (57) we then have for $\forall j = 0, 1, 2, \cdots$

$$\begin{aligned} \widehat{v_{j+1}}^{\lambda} &\approx \lambda\, \widehat{v_j}^{\lambda} \\ \widehat{(v_{j+1} \otimes v_{j+1})}^{\lambda} &\approx \lambda\, \widehat{(v_j \otimes v_j)}^{\lambda} \\ \widehat{p_{j+1}}^{\lambda} &\approx \lambda\, \widehat{p_j}^{\lambda} \end{aligned} \tag{79}$$

## 7 Remarks

Having a decomposition of the real iterated values $g_k$ as

$$\mathbb{N}_0 \ni k \mapsto g_k = \sum_{l=1}^{p} v_l\, {\lambda_l}^k \tag{80}$$

with $|\lambda_l| \leq 1$ then with $(\lambda_l, v_l)$ appear also the conjugated elements $\left(\overline{\lambda_l}, \overline{v_l}\right)$. So the decomposition can be written as sum of terms $2\,\mathrm{Re}\, v_l\, {\lambda_l}^k$. These terms could be understood as vectors in a two dimensional subspace moving with the time step $k$. They can be animated in this form.

Eigenvalues with modulus lower than 1 belong to eigenmodes, which disappear during ongoing iterations. Even for eigenvalues with modulus near to 1, e.g. $|\lambda_l| = 0.999$ the term $v_l\,\lambda_l^{\,k}$ is reduced by a factor of 0.37 for $k = 1000$.

Changing the degree $p$ of the polynom coefficient vector $c$ has influence on some eigenvalues but not on all. The latter might be candidates for determining the continuous part of the spectrum. On the one hand the degree $p$ of $c$ should be small to limit the number of modes; on the other hand a small degree enlarges the approximation error $\mu$. There must be a balance between the degree $p$ of $c$, this is the number of eigenvalues, and $m$ as number to take the mean of submatrices. The sum $m + p$ is the number of given measurements. The continuous part of the spectrum is not addressed by this approach. It might be that it numerically shows up by the eigenvalues with eigenvectors with a small norm which are locally uniformly distributed. Our example gives an impression.

For the largest possible $p = n$, we have $\mathfrak{A}(c) = c$ and $(c, \rho)$ can be an eigenpair for smallest eigenvalue of $H$. This is the setting for the Dynamic Mode Decomposition (DMD) of [12].

## 8 Computational Costs and Performance Aspects

An essential part of the computational effort is the calculation of the product $G^T G$ in (8) resp. $\widehat{G}^T\widehat{G}$ in (65).

The number of columns of the matrix $G$ is the number of used time steps. The number of rows of $G$ is typically very large. In case of a discretized PDE this number is #(variables per node)$*$#(discretization nodes) $*$# (repetitions of the sequence). The effort for reading might be large. Assume a discretization grid with $10^9$ nodes with 4 variables per node with 8 B and $n = 1000$ time steps. The stored data are $10^9 * 4 * 1000 * 8$ B = 29.1 TB. A system with a effective read bandwidth of 100 GB/sec would need 298 s to read the data.

The number of operations for the calculation of $G^T G$ is $2 * 4 * 10^9 * 1000^2$ Flop= $8 * 10^{15}$ Flop= $8 * 10^3$ TFlop. On a system running with an effective performance of 10 TFlops we need 800 s computing time which is comparable with the read time. The computing time increases with second order of the number of time steps whereas the reading the input data with the first order. That means, that the process is computationally limited. The second compute intensive part is the multiplication of the matrix $G$ with the matrix consisting on the pseudoeigenvectors $(w_l)_{l=1,\cdots,p}$. The needed effort is comparable to the effort for the calculation of $G^T G$.

Opposite to the size of $G$ the matrix $G^T G$ typically has a relatively small size given by the number of intermediate time steps.

A generalized eigenproblem of size $n - p$ has to be solved iteratively for determining $c$ with degree $p$. This takes relevant time, if $p$ is relatively small and many iterations have to be made.

## 9 Application Example

As example we show the simulation of the flow in an artery near the heart aortic-valve with OpenFOAM. The geometry was taken from MRI-data of Fraunhofer Mevis [9] (https://www.mevis.fraunhofer.de/) as well as the estimations of inflow and outflow conditions. The simulation has been done by [11]. Figures 2 and 3 show the square root of the norm of the eigenmodes over the phase of their eigenvalues for time step 0 and time step 2457. The second figure shows disappearing modes with modulus smaller 1 . The distribution also shows prominent eigenvalues which are nearly multiples of a smallest. This property is expected, because $\lambda$ being a Koopman eigenvalue, also $\lambda^m$ for $m \in \mathbb{N}$ is an Koopman eigenvalue.

Figures 4, 5, 6, and 7 show the the vector field $2\,\mathrm{Re}\, v_l\ \lambda_l{}^k$ of the eigenmode $l$ with the second largest norm in the upper part of the aorta for the time steps

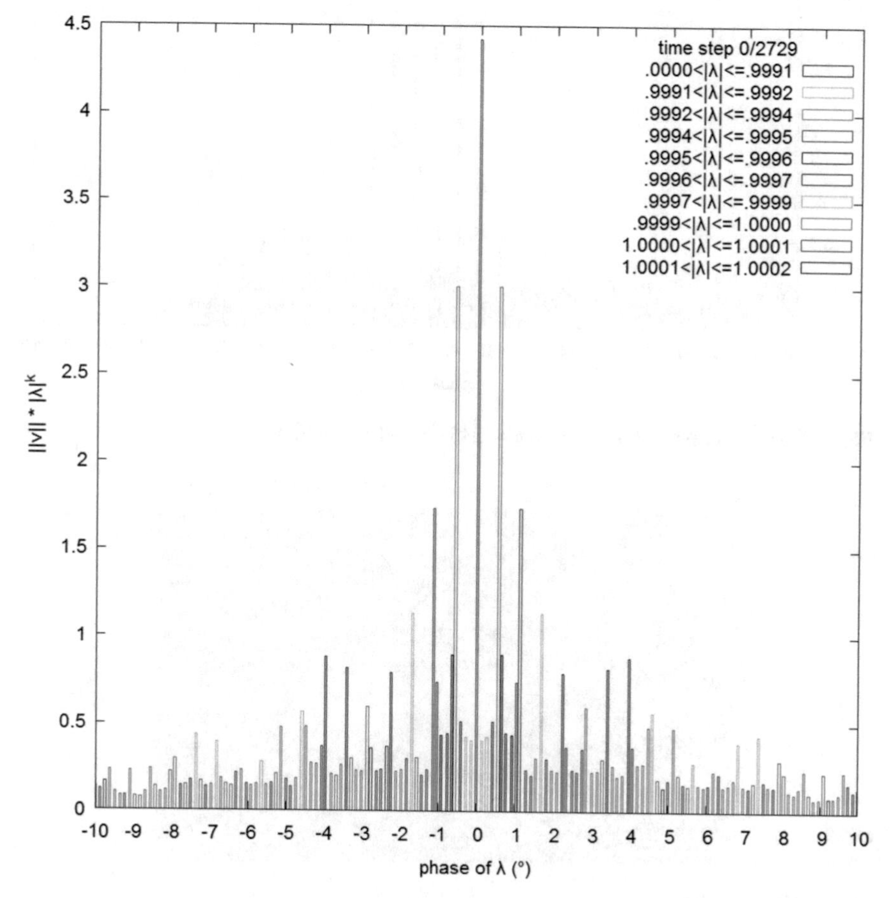


**Fig. 2** $\|v_l \lambda_l^k\|$ over phase for time step $k = 0$ in $[-10^\circ : +10^\circ]$

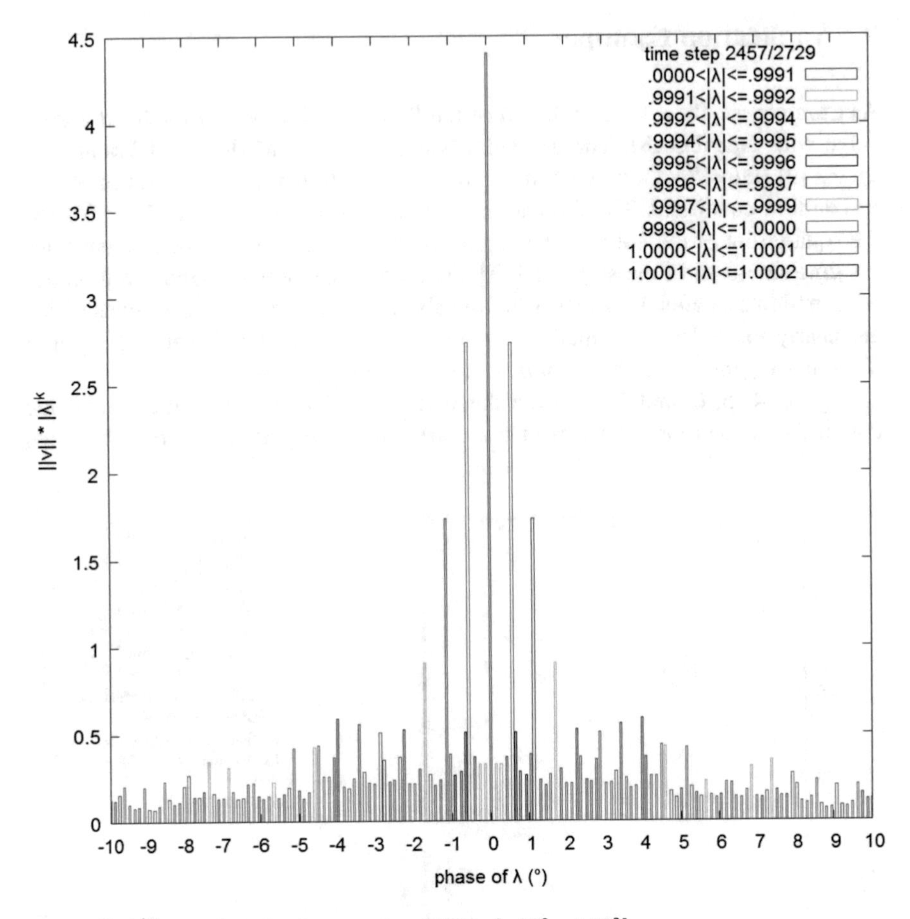


**Fig. 3** $\|v_l \lambda_l^k\|$ over phase for time step $k = 2457$ in $[-10° : +10°]$

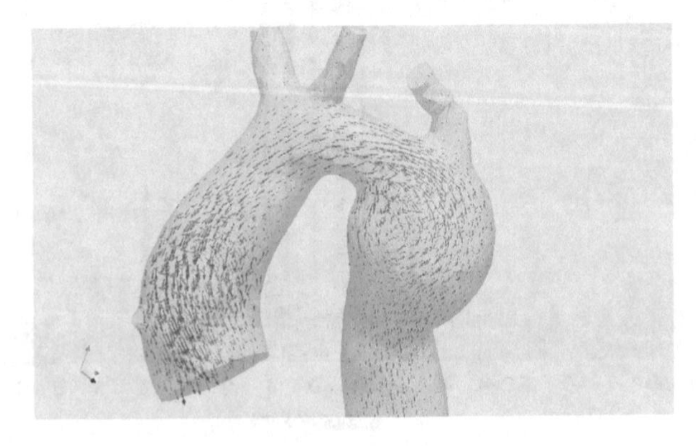

**Fig. 4** Time step 661

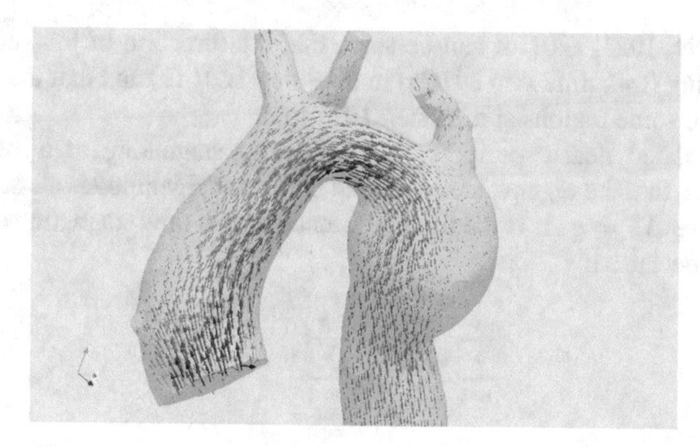

**Fig. 5** Time step 841

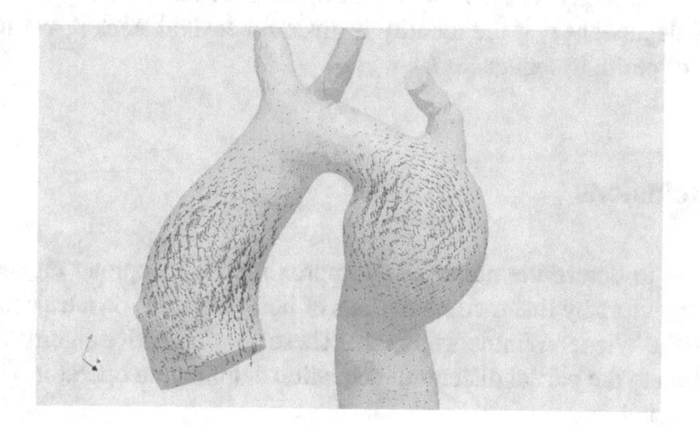

**Fig. 6** Time step 1021

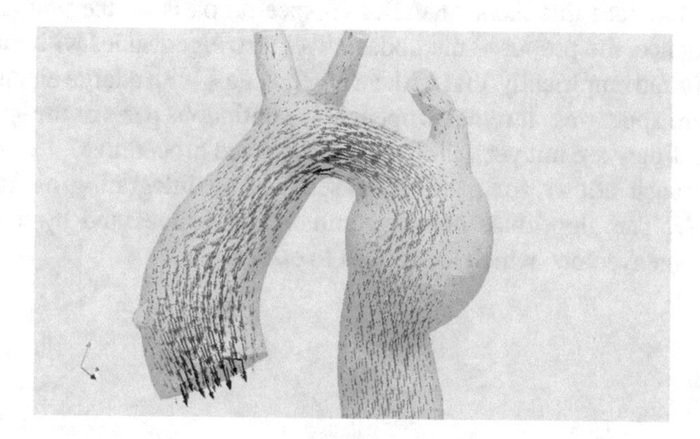

**Fig. 7** Time step 1201

$k = 661, 841, 1021, 1201$. It can be seen, that the direction of the vector field is reversed going from time step 841 (5) to time step 1201 (7) and that the vector field disappears in some regions at time step 1021 (6).

The simulated nearly periodic flow is nearly stagnating at a certain time. Interesting is, that the eigenvalues $\lambda_l$ of the dominant eigenmodes all nearly satisfy $\lambda_l^q = 1$ where $\Delta T = q\,\Delta t$ is the time difference of two flow stagnation events. This is expected because if

$$\sum_{l=1}^{p} v_l\,\lambda_l^{k} = \sum_{l=1}^{p} v_l\,\lambda_l^{k+q} \tag{81}$$

it is clear that $\sum_{l=1}^{p} v_l \left(\lambda_l^{k+q} - \lambda_l^{k}\right) = 0$ and for linearly independent $v_l$ that also $\lambda_l^{k+q} - \lambda_l^{k} = 0$ implying for $\lambda_l \neq 0$ that $\lambda_l^q = 1$ for all $l$. This remains true even for linearly dependent $v_l$ if the identity is given for several steps $k$. We see here the peculiarity of periodic sequences $k \mapsto g_k$.

## 10 Conclusions

It is possible to determine numerically approximative Koopman eigenvectors for nonlinear operators by linear combinations of iterated values on a trajectory. This is in line with the Wiener-Wintners theorem. These approximative eigenvectors have a direct relation to the partial differential equation defining the operator. The physical meaning is not clear.

We have shown the relationship of variants of the Dynamic Mode Decomposition to Fourier analysis using the theorem of Herglotz-Bochner for the eigenvalues with modulus 1. Whereas this theory handles on spectral parts on the unit circle for an infinite sequence, the proposed methods deliver also eigenvalues with modulus less than 1. We found empirically, that with an eigenvalue $\lambda$ with a large eigenvector also $\lambda^2, \lambda^3, \cdots$ are appearing. It remains open, how continuous parts of the spectrum can be handled, if they are not yet visible by the proposed procedures.

The approach allows for handling ensembles by integrating all members in the matrix $G$. The algorithms deliver common eigenvalues and by the described procedures eigenvectors, which are related to each other.

## References

1. Bellow, A., Losert, V.: The weighted pointwise ergodic theorem and the individual ergodic theorem along subsequences. Trans. Am. Math. Soc. **288**(1), 307–345 (1985). https://doi.org/10.1090/S0002-9947-1985-0773063-8

2. Besicovitch, A.S.: On generalized almost periodic functions. Proc. Lond. Math. Soc. **s2-25**, 495–512 (1926). doi:10.1112/plms/s2-25.1.495
3. Budišić, M., Mohr, R., Mezić, I.: Applied Koopmanism. Chaos **22**, 047510 (2012). doi:10.1063/1.4772195. http://dx.doi.org/10.1063/1.4772195
4. Chen, K.K., Tu, J.H., Rowley, C.W.: Variants of dynamic mode decomposition: boundary condition, Koopman, and Fourier analyse. J. Nonlinear Sci. **22**(6), 887–915 (2012)
5. Eisner, T., Farkas, B., Haase, M., Nagel, R.: Operator Theoretic Aspects of Ergodic Theory. Graduate Texts in Mathematics. Springer, Cham (2015)
6. Koopman, B.O.: Hamiltonian systems and transformations in Hilbert space. Proc. Natl. Acad. Sci. U. S. A. **17**(5), 315–318 (1931)
7. Küster, U.: The spectral structure of a nonlinear operator and its approximation. In: Sustained Simulation Performance 2015: Proceedings of the Joint Workshop on Sustained Simulation Performance, University of Stuttgart (HLRS) and Tohoku University, pp. 109–123. Springer, Cham (2015) ISBN:978-3-319-20340-9. doi:10.1007/978-3-319-20340-9_9
8. Küster, U.: The spectral structure of a nonlinear operator and its approximation II. In: Sustained Simulation Performance 2016: Proceedings of the Joint Workshop on Sustained Simulation Performance, University of Stuttgart (HLRS) and Tohoku University (2016). ISBN:978-3-319-46735-1
9. Mirzaee, H., Henn, T., Krause, M. J., Goubergrits, L., Schumann, C., Neugebauer, M., Kuehne, T., Preusser, T., Hennemuth, A.: MRI-based computational hemodynamics in patients with aortic coarctation using the lattice Boltzmann methods: clinical validation study. J. Magn. Reson. Imaging **45**(1), 139–146 (2016). doi:10.1002/jmri.25366
10. Rellich, F.: Störungstheorie der Spektralzerlegung I., Analytische Störung der isolierten Punkteigenwerte eines beschränkten Operators. Math. Ann. **113**, 600–619 (1937)
11. Ruopp, A., Schneider, R., MRI-based computational hemodynamics in patients. In: Resch, M.M., Bez, W., Focht, E. (eds.) Sustained Simulation Performance 2017 (abbrev. WSSP 2017). Springer, Cham (2017). doi:10.1007/978-3-319-66896-3
12. Schmid, P.J.: Dynamic mode decomposition of numerical and experimental data. J. Fluid Mech. **656**, 24 (2010)
13. Schröder, E.: Ueber iterirte Functionen. Math. Ann. **3**(2), 296–322 (1870). doi:10.1007/BF01443992

# Part III
# Optimisation and Vectorisation

# Code Modernization Tools for Assisting Users in Migrating to Future Generations of Supercomputers

**Ritu Arora and Lars Koesterke**

**Abstract** Usually, scientific applications outlive the lifespan of the High Performance Computing (HPC) systems for which they are initially developed. The innovations in the HPC systems' hardware and parallel programming standards drive the modernization of HPC applications so that they continue being performant. While such code modernization efforts may not be challenging for HPC experts and well-funded research groups, many domain-experts and students may find it challenging to adapt their applications for the latest HPC systems due to lack of expertise, time, and funds. The challenges of such domain-experts and students can be mitigated by providing them high-level tools for code modernization and migration. A brief overview of two such high-level tools is presented in this chapter. These tools support the code modernization and migration efforts by assisting users in parallelizing their applications and porting them to HPC systems with high-bandwidth memory. The tools are named as: Interactive Parallelization Tool (IPT) and Interactive Code Adaptation Tool (ICAT). Such high-level tools not only improve the productivity of their users and the performance of the applications but they also improve the utilization of HPC resources.

## 1 Introduction

High Performance Computing (HPC) systems are constantly evolving to support computational workloads at low cost and power consumption. While the computing density per processor has increased in the last several years, the clock speed of the processors has stopped increasing to prevent unmanageable increase in the temperature of the processor, and to limit the gap between the speed of the processor and the memory. This trend has resulted in HPC systems that are equipped with manycore processors and deep memory hierarchies. To achieve high-performance on such HPC systems, appropriate level of parallelization, vectorization, and memory optimization are critical.

---

R. Arora (✉) • L. Koesterke
Texas Advanced Computing Center, The University of Texas at Austin, Austin, TX, USA
e-mail: rauta@tacc.utexas.edu; lars@tacc.utexas.edu



M.M. Resch et al. (eds.), *Sustained Simulation Performance 2017*,
DOI 10.1007/978-3-319-66896-3_4

As a sample of evolution in the HPC landscape, consider the three flagship HPC systems provisioned by the Texas Advanced Computing Center (TACC) in the last 10 years: Ranger, Stampede, and Stampede2. Ranger debuted in the year 2008. It was equipped with AMD Opteron quad core processors, delivered approximately 579 TFLOPs of peak performance, and was in production for about 5 years. Stampede debuted in the year 2013 and is still in production. It can deliver approximately 10 PFLOPs of theoretical peak performance, and is equipped with Intel Sandy Bridge processors, Intel Knight's Corner coprocessors, and Nvidia K20 GPUs. The Stampede2 system, that is currently under development, will be equipped with the Intel Haswell processors, future generation Intel Xeon processors, and Intel Knight's Landing (KNL) processors. The estimated theoretical peak performance of Stampede2 is about 18 PFLOPs. Figure 1 depicts the rate of evolution in the HPC landscape considering the aforementioned systems deployed by TACC as examples.

For migrating applications from a system like Ranger to Stampede, and from Stampede to Stampede2, some level of software reengineering may be required to enhance the performance of the applications. The required reengineering may be in the form of increasing the level of parallelization in the applications by incorporating both OpenMP and MPI programming paradigms, or improving code vectorization to effectively take advantage of the Intel Sandy Bridge and Knight's Corner coprocessors, or optimizing the memory access pattern of the applications to benefit from the High-Bandwidth Memory (HBM) on the KNL processors.

As evident from the example of TACC resources, even though the current and future generation HPC systems may be equipped with high-end hardware components, they may not offer the same range of diversity in processing elements as compared to the previous generation systems. For example, unlike the Stampede system, the Stampede2 system does not include GPUs and Intel Knight's Corner

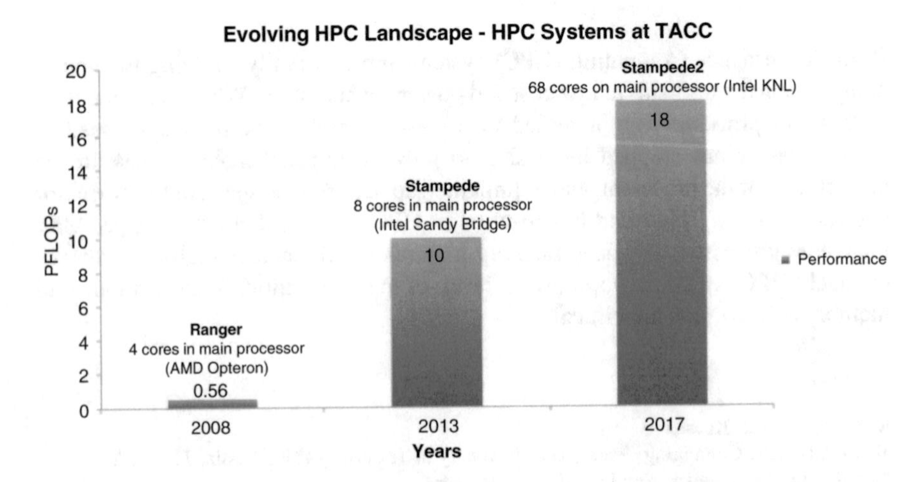


**Fig. 1** Evolution in the HPC landscape

(KNC) coprocessors. Thus, the applications written using CUDA to take advantage of the GPUs, or offload programming model for KNC coprocessors will not be able to run on Stampede2. This implies that the HPC applications may need to be updated or reengineered at the same frequency as the systems on which they are supposed to run (typically, 4 years on an average).

Together with the evolution in the HPC hardware over the last several years, the parallel programming standards have also been continuously evolving by including new features for improved performance on modern HPC systems. For example, the nonblocking collective calls were added to the MPI 3.0 standard [1], and the `taskloop` construct was added to the OpenMP 4.5 standard [2]. Incorporation of such new features in the existing HPC applications requires time and effort in climbing the learning curve, and in reengineering the applications.

Given the discussion presented thus far, the questions that arise are:

- Are domain experts ready to invest time and effort in continuously modernizing their applications?
- Do we have enough trained workforce to support HPC code modernization and migration efforts?

During the process of pursuing the answers to the aforementioned questions, we found that the domain experts prefer to spend time in doing science rather than keeping up with the evolution in the HPC hardware and parallel programming standards. We also found that there is a shortfall of trained workforce in the area of parallel programming. Therefore, we need high-level tools for assisting users in their HPC code modernization and migration efforts. To address this need, we are developing high-level tools for supporting (1) the parallelization of serial applications using MPI, OpenMP, CUDA, and hybrid programming models, and (2) the migration of applications to the KNL processors. These tools are named as Interactive Parallelization Tool (IPT) and Interactive Code Adaptation Tool (ICAT). Before we present an overview of these tools, we explain the typical process of manual code modernization and migration, and share our perspective on the automation or semi-automation of the process. We also present a short overview of the KNL architecture and discuss some key considerations for porting applications to this architecture.

## 2 General Process for Manual Code Modernization and Migration

The traditional process of manually upgrading the code for taking advantage of the latest HPC systems is as follows:

1. Learn about a new hardware feature
2. Gauge the applicability of the hardware and the potential reward from the analysis of current bottlenecks (code profiling) and theoretical considerations

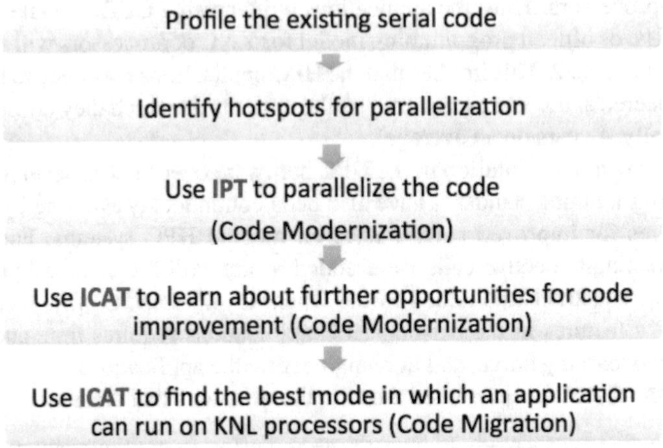

Profile the existing serial code

Identify hotspots for parallelization

Use IPT to parallelize the code (Code Modernization)

Use ICAT to learn about further opportunities for code improvement (Code Modernization)

Use ICAT to find the best mode in which an application can run on KNL processors (Code Migration)

**Fig. 2** Using IPT and ICAT for code modernization

3. Develop a general concept of the necessary code modifications
4. Learn the syntax of the new programming model or the interface
5. Write a toy code
6. Insert pieces of the toy code into the target code for exploration and testing
7. Estimate performance gain (based on tests not on theoretical considerations) and judge the quality of the implementation
8. Modify, test, and release production code
9. Experiment with the various runtime options and environment variables
10. Learn good coding practices

Outlined above is a typical process and not all users go through all the steps in this process. The high-level tools that we are developing—IPT and ICAT—will assist the users in the aforementioned process, and in doing so, will significantly speed up many of its steps. IPT can assist the users in steps 1, 3–6, 8 and 10 mentioned above. ICAT can assist the users in steps 1 to 6, and 9–10. Figure 2 summarizes the steps for using IPT and ICAT together during the process of modernizing and migrating applications written in C, C++ and Fortran base languages.

## 3 Using IPT for Code Modernization (Parallelization)

IPT semi-automates the parallelization of existing C/C++ applications, and by doing so, helps in running the applications optimally on the latest HPC systems [1, 2]. It can support the parallelization of applications using any of the following parallel programming models: Message Passing Interface (MPI) [3], OpenMP [4], and CUDA [5]. IPT uses the specifications for parallelization as provided by the users (i.e., what to parallelize and where) and its knowledgebase of parallel programming

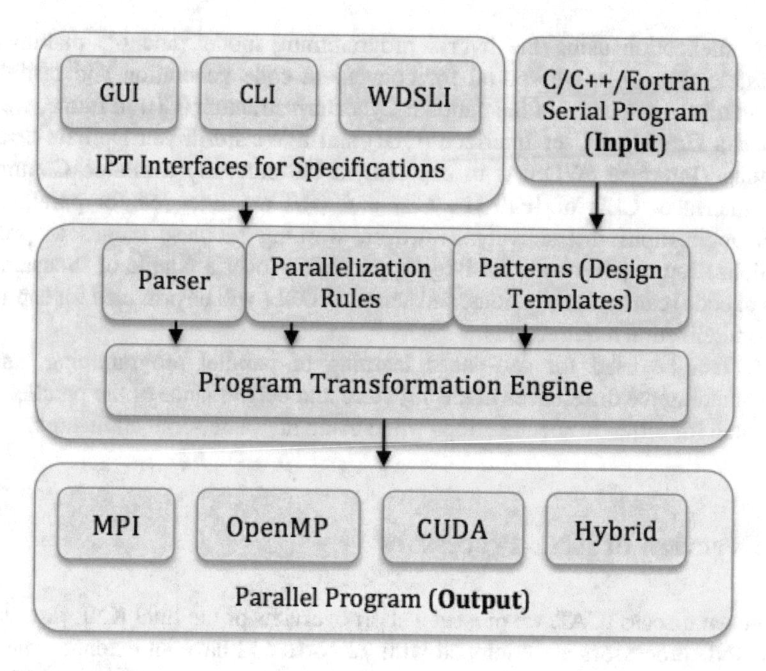


**Fig. 3** Overview of IPT

expertise (encapsulated as design templates and rules), to generate parallel versions of the serial programs.

A high-level overview of the process of serial to parallel code transformation using IPT is shown in Fig. 3. As shown in Fig. 3, the user provides the serial code as input to IPT. This serial code is parsed by IPT, and as a next step, IPT prompts the user for additional input (or specifications). The user chooses the desired parallel programming model (MPI/OpenMP/CUDA) for their output code, applicable design patterns (e.g., stencil and pipeline), and hotspots for parallelization. The user also provides additional information when prompted by IPT for variable dependency analysis of the input serial application.

Using the infrastructure provided by a Program Transformation Engine (PTE) named ROSE [6], and user input, IPT analyzes and transforms the serial code into a parallel one. As required, it inserts, deletes and updates the program statements in the serial code for generating the parallel code. The generated parallel code is accessible to the user and is well annotated to give insights into the parallelization process. The design templates in IPT contain rules for parallelization and patterns for supporting data movement (e.g., data distribution, data collection, and data exchange) in MPI programs. During parallelization, IPT weaves these design templates into the serial code by the means of appropriate function calls.

We are currently working on (1) extending the capabilities of IPT for parallelizing additional categories of C/C++ applications (e.g., divide-and-conquer), (2) prototyping support for parallelizing Fortran applications, (3) adding support for parallelization using the hybrid programming model, and (4) making IPT accessible through a web-portal for convenient code generation and testing on computational resources of the national CyberInfrastructure (CI). In future, we will support a Graphical User Interface (GUI) and a Wizard-driven Domain-Specific Language Interface (WDSLI) to supplement the currently available Command-Line Interface (CLI) of IPT. The CLI and GUI are intended for parallelizing small applications interactively. However, working in these modes to provide parallelization requirements for large applications (over a couple of thousands of lines of code) can become tedious, and hence, WDSLI will be provided for capturing the parallelization requirements.

IPT can be used for self-paced learning of parallel programming, and in understanding the differences in the structure and performance of the parallel code generated for different specifications while using the same serial application.

# 4 Overview of KNL Processors

Before we discuss ICAT, we present a short overview of the Intel KNL processors. Intel KNL processors are equipped with 72 cores and have an extended memory architecture. The cores on these processors are organized in 36 pairs and each pair is known as a tile. These processors have a 16 GB High-Bandwidth Memory (HBM) called Multi-Channel DRAM (MCDRAM), alongside the traditional DDR4 memory that is approximately 400 GB [7].

## *4.1 Multiple Memory Modes*

The MCDRAM can be configured for use in three different memory modes:

1. Cache mode: As a third-level cache that is under the control of the run-time system,
2. Flat mode: As an addressable memory like DDR4 that is under user control, or
3. Hybrid mode: Part of MCDRAM is configured in cache mode and part of it is configured in flat mode.

The cache mode is more convenient to use because it does not require any code modification or user interaction and ensures high performance for applications that have a small memory footprint. The applications having large memory footprints are likely to see a drop in their performance if they are run in cache mode due to frequent cache misses. It may be advantageous for such applications to manage the cache from within the code and to store only specific arrays in the MCDRAM—that is, using flat mode is recommended for such applications.

For selectively allocating arrays on MCDRAM, the existing code for dynamic memory allocation is modified to use special library calls or directives that are available through the HBWMALLOC interface [8]. In the case of C/C++ applications, the function calls for dynamic memory allocation—`calloc`, `malloc`, `realloc`, and `free` functions—are replaced with the analogous functions in the HBWMALLOC interface—`hbw_calloc`, `hbw_malloc`, `hbw_realloc`, and `hbw_free` functions. A header file for HBWMALLOC interface is also included. For allocating memory from MCDRAM in Fortran applications, a directive with the FASTMEM attribute is added after the declaration of the allocatable data structure of interest.

Understanding the concept of two memories (MCDRAM and DDR4), and doing the required code modifications for using MCDRAM effectively may not be difficult by itself. However, it takes time to understand the syntax of the code required for memory allocation on MCDRAM, to learn about the additional tools for understanding the application characteristics (cache miss or hit rate, sizes of memory objects etc.), and more importantly to derive the logic of the decision tree for allocating memory on MCDRAM or DDR4. In order to develop a portable code, it is also important to implement appropriate logic for handling situations that can give rise to runtime errors. For example, the code should handle situations where the user attempts to dynamically allocate more than 16 GB of memory on MCDRAM or tries to run the code on processors that do not support MCDRAM.

### *4.2 Multiple Cluster Modes*

The tiles on a KNL processor are connected to each other with a mesh interconnect. Each core in a tile has its own L1 cache and a 1 MB L2 cache shared with the other core. The L2 cache on all the tiles are kept coherent with the help of a Distributed Tag Directory (DTD), organized as a set of per-tile Tag Directories (TDs). The TDs help in identifying the location and the state of cache lines on-die. When a memory request originates from a core, an appropriate TD handles it, and if needed passes the request to the right memory controller. All on-die communication for handling such memory requests happens over the mesh interconnect. To achieve low latency and high bandwidth of communication with caches, it is important that the on-die communication is kept as local as possible. For handling this on-die communication optimally, KNL processors can be configured in different cluster modes:

1. All-to-All: The memory addresses are uniformly distributed across all TDs, and this mode is used mainly for troubleshooting purposes or when other modes cannot be used because it can result in high latency for various on-die communication scenarios.
2. Quadrant or Hemisphere: The tiles on a processor are virtually divided into four parts called quadrants, and each quadrant is in proximity to a memory controller. The memory addresses controlled by the memory controller in a quadrant are mapped locally to the TD in that quadrant. This arrangement reduces the latency of a cache miss as compared to the all-to-all mode because the memory controller

and TD's are in the same locality. However, the TD servicing the memory request may not be local to the tile whose core initiated the memory request. The hemisphere mode is similar to the quadrant mode with the difference that the tiles on the chip are divided into two parts instead of four.
3. Sub-NUMA (SNC-4/SNC-2): Similar to the quadrant mode, the tiles are divided into four (SNC-4) or two (SNC-2) parts in this mode too. However, unlike in the quadrant mode, in the sub-NUMA mode, each part acts as a separate NUMA node. This means that, the core requesting memory access, the TD, and the memory channel for servicing the memory access request, are all in the same part.

While the quadrant mode could work well for majority of the applications, the sub-NUMA mode can result in better performance for multi-threaded NUMA-aware applications by pinning the threads and memory to the specific quadrants or hemispheres on each NUMA node. However, the users may have to do their own testing to find out the best cluster mode and the runtime options for their applications.

## 5 Using ICAT for Code Modernization and Migration (Porting Code to KNL Processors)

ICAT can assist users in porting their applications to KNL processors by helping them select the best memory mode and cluster mode and suggesting runtime options. It can reengineer their application code also to optimally take advantage of the MCDRAM while keeping it portable enough to run on other systems that do not support MCDRAM.

By using ICAT, a user can very quickly understand source code modifications, potential performance gains, and learn good coding practices for porting their applications to the KNL nodes. They can then move on to modifying their real application code using ICAT itself, or may cut-and-paste boilerplate code generated by ICAT. Thus, ICAT offers three key benefits to the users: (a) enables users to make a decision quickly regarding the best memory mode and cluster mode for their applications, (b) teaches good coding practices, and (c) assists in changing production code.

Figure 4 shows an overview of the functioning of ICAT, which is invoked from the command-line. ICAT prompts the user for input, such as, the name of an application's executable, path to the executable, and path to the application's source code. The user selects an appropriate advisor mode in which ICAT can run. The available advisor modes are: memory mode advisor, code adaptation advisor, cluster mode advisor, advanced vectorization advisor, and memory optimization advisor. ICAT performs memory usage and performance analyses by running the executable provided by the user with `perf` [9], and then if needed, with Vtune [10]. Metrics are also collected from the processes associated with the executable while it is

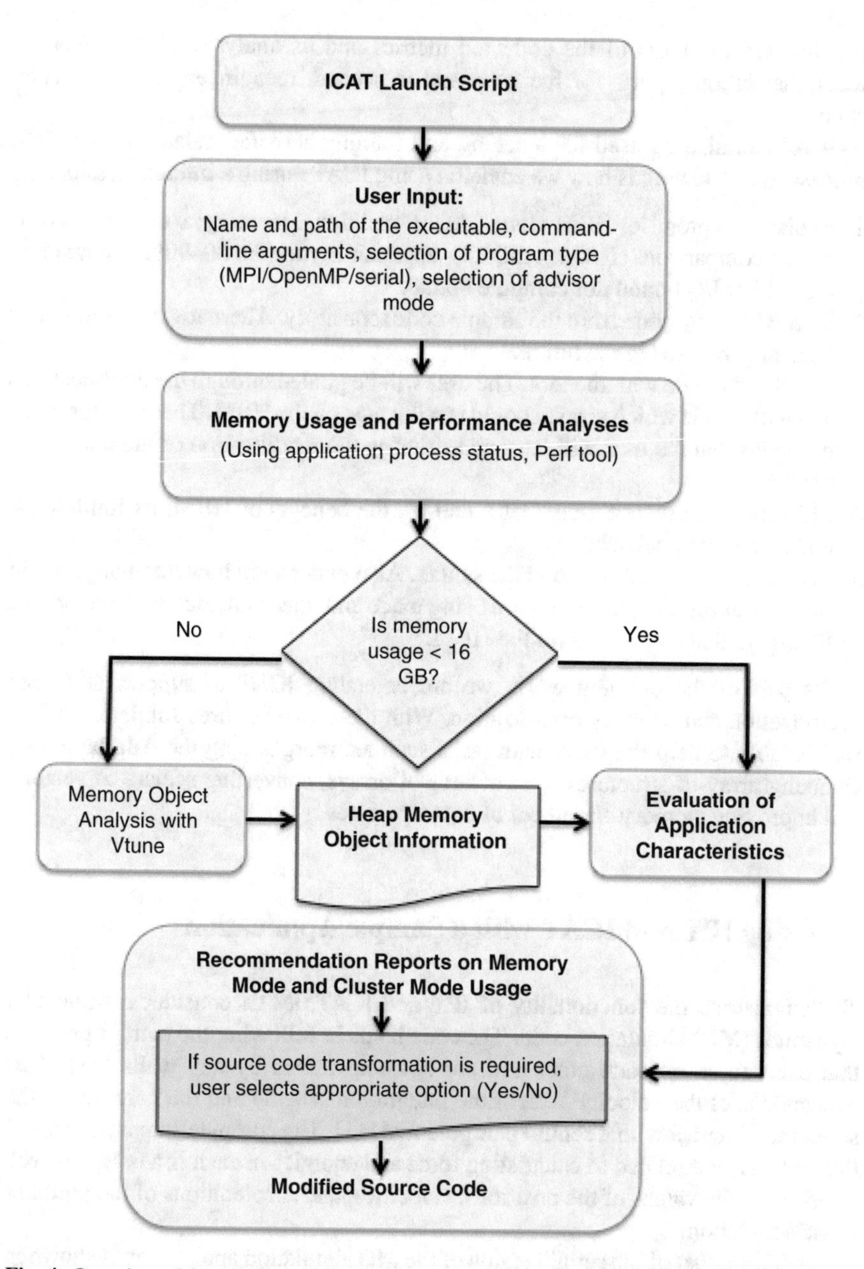


**Fig. 4** Overview of functioning of ICAT

running. On the basis of the collected metrics and its analyses, ICAT generates recommendation reports for the user, and if needed, reengineers the application code.

ICAT can also be used for teaching and training activities related to the KNL processors. Following is how we envision using ICAT during a training session:

1. Explain the premise: HBM alongside the traditional memory; some raw performance comparisons (HBM v. DDR4); applicability for the 80–90% of cases that are neither I/O bound nor compute bound.
2. Start with a toy code from the sample code repository. Alternatively a small user code may be used in the future.
3. Run the toy code with the tool. The user will be guided through the modifications and will decide which arrays should be allocated on the HBM. These are the same decisions that the user will later make during the modification of the real-world applications.
4. Measure performance gain. Get a feel for the benefits of HBM, its limitations, and potential drawbacks.
5. Inspect the modified code and the syntax. Also understand how internally, i.e., in the code at runtime, decisions are being made and how a fallback is implemented for arrays that are too big for the HBM.

As part of the ongoing work, we are extending ICAT to support advanced vectorization and memory optimization. With these two features supported, ICAT will be able to help the users with tasks such as: reorganizing the data layout by changing array-of-structures to structures-of-arrays, converting scalars to vectors, and improving memory alignment of data structures.

## 6 Using IPT and ICAT with a Sample Application

To demonstrate the functionality of IPT and ICAT, let us consider a Molecular Dynamics (MD) simulation code. The code helps in following the path of particles that exert force on each other and are not constrained by any walls [11]. This MD code uses the velocity Verlet time integration scheme and the particles in the simulation interact with a central pair potential [11]. The compute-intensive steps in this test case are related to calculating force and energies in each time-step, as well as updating the values of the positions, velocities, and accelerations of the particles in the simulation.

A code snippet of the serial version of the MD simulation application is shown in Fig. 5 and the complete code can be accessed at [12]. In order to parallelize this code, the computations in the for-loop beginning at line # 3 of Fig. 5 should be distributed across multiple threads or processes. The values of the kinetic energy (`ke`) and potential energy (`pe`) are augmented in every iteration of this for-loop. Therefore, with the distribution of the iterations of the for-loop, only the partial values of `ke` and `pe` will be computed by each thread or process. Hence, all the partial values of

```
1. void compute ( …,double f[],double *pot,double *kin ){
2.      //other code
3.      for ( k = 0; k < np; k++ ){
4.        // Compute the potential energy and forces.
5.        //other code
6.        for ( j = 0; j < np; j++ ){
7.          if ( k != j ){
8.            d = dist ( nd, pos+k*nd, pos+j*nd, rij );
9.            //other code
10.              pe = pe + 0.5 * pow ( sin ( d2 ), 2 );
11.         for ( i = 0; i < nd; i++ ){
12.           f[i+k*nd]=f[i+k*nd] - rij[i]*sin (2.0 * d2)/d;
13.         }}}
14.    // Compute the kinetic energy
15.    for ( i = 0; i < nd; i++ ) {
16.          ke = ke + vel[i+k*nd] * vel[i+k*nd];
17.    }}
18.    ke = ke * 0.5 * mass;
19.    *pot = pe;
20.    *kin = ke;
21.    return;
22.    }
```

**Fig. 5** Code snippet—MD simulation, serial version

`ke` and `pe` computed using the multiple independent threads and processes should be combined meaningfully to obtain accurate results. For combining the partial values of `ke` and `pe`, a reduction operation is needed.

## *6.1 Using IPT to Parallelize the MD Application*

In this section, we will demonstrate the usage of IPT by generating an OpenMP version of the serial MD simulation application. As shown in Fig. 6, IPT is invoked from the command-line, and the path to the file containing the serial code is provided. Next, a parallel programming model is selected, here OpenMP. This is followed by selecting the function that contains the hotspot for parallelization from the list of functions presented by IPT.

IPT analyzes the code in the function selected by the user (see Fig. 6), and presents a list of the for-loops that can be parallelized (because, in this example, the user chose to parallelize for-loops). As shown in Fig. 7, users can accept or decline to parallelize the for-loops presented by IPT for parallelization.

For constructing the clauses of the OpenMP directives, IPT can detect the variables that should be part of the `shared`, `private`, and `firstprivate` clauses. However, IPT relies on the user-guidance for constructing the reduction clause of the relevant OpenMP directives (`#pragma omp parallel`, or `#pragma omp parallel for`, or `#pragma omp for` ). Reduction variables are the variables

```
$ ./IPT md.c

Please select a parallel programming model from the following available
options:
1. MPI
2. OpenMP
3. CUDA
2

Would you like to parallelize a for-loop?(Y/N)
y

Please choose the function that you want to parallelize from the list
below
1 : main
2 : compute
3 : cpu_time
4 : dist
5 : initialize
6 : r8mat_uniform_ab
7 : timestamp
8 : update
2
```

**Fig. 6** Invoking IPT for parallelizing MD simulation application

that should be updated by the OpenMP threads by creating a private copy for each reduction variable and initializing them for each thread. The values of the variables from each thread are combined according to a mathematical operation like sum, multiplication, etc. and the final result is written to a global shared variable. IPT generates a list of potential variables that can be part of a reduction clause, and prompts the user to select the relevant variables and appropriate reduction operation.

In some cases, IPT is unable to analyze the pattern related to updating the array elements at the hotspot for parallelization. This typically happens when multiple levels of indirection are involved during the process of updating the values of the array elements. In such situations, as shown in Fig. 8, IPT relies on the user to provide additional information on the nature of the update operation on their array elements.

IPT also prompts the user to confirm if the I/O in their application should be happening from a single thread (or process) or using all the threads (or processes involved in the computation). If there is any region of code that should not be executed in parallel, then, the user can inform IPT about this as well.

A snippet of the parallelized version of the MD simulation application is shown in Fig. 9. In addition to updating the code at the hotspot for parallelization by inserting OpenMP directives (lines 3–5 of Fig. 9), IPT inserts appropriate library header files as well.

```
    for ( k = 0; k < np; k++ ){
        // Compute the potential energy and forces.
        //other code
        for ( j = 0; j < np; j++ ){
          if ( k != j ){
             //other code
          }}
        // Compute the kinetic energy
        for ( i = 0; i < nd; i++ ) {
          ke = ke + vel[i+k*nd] * vel[i+k*nd];
        }
    }

Is this the for loop you are looking for?(y/n)
y

Reduction variables are the variables that should be updated by the
OpenMP threads by … Below is the list of potential reduction variables
in the region of code selected for parallelization:

1. nd type is int
2. k type is int
…
7. pe type is double
8. ke type is double

How many variables in the listed above should be selected as reduction
variables? If there are no reduction variables, please enter 0.
2

Please enter a number corresponding to the reduction variable in the
list above.
7

Please select the type of reduction operation for the selected
variable:
1. Addition
2. Subtraction
3. Min
4. Max
5. Multiplication
1

…
```

**Fig. 7** User guiding IPT in selecting the reduction variables

## 6.2 *Using ICAT to Adapt the MD Application for KNL Processors*

As described in Sect. 3, before running an application on KNL processors, it is important to understand the application's characteristics so that the best memory mode and the cluster mode configuration of the KNL processors can be selected for it. Depending upon the memory needs of the application, some reengineering may also be required for selectively allocating memory for specific arrays on MCDRAM.

```
–
IPT is unable to perform the dependency analysis of the array named [
rij ] in the region of code that you wish to parallelize. Please enter
1 if the entire array is being updated in a single iteration of the
loop that you selected for parallelization, or, enter 2 otherwise.
1

Are there any lines of code that you would like to run either using a
single thread at a time (hence, one thread after another), or using
only one thread?(Y/N)
n

Would you like to parallelize another loop?(Y/N)
n

Are you writing/printing anything from the parallelized region of the
code?(Y/N)
n

Running Consistency Tests
```

**Fig. 8** User guiding IPT in analyzing the nature of the updates made to the array values

```
1. void compute ( …,double f[],double *pot,double *kin ){
2.      //other code
3. #pragma omp parallel default(none) shared(pe,ke,np,f,pos,vel,nd,PI2)
   private(k,i,j,d,d2) firstprivate(rij)
4. {
5. #pragma omp for reduction ( + :pe,ke)
6. for (k = 0; k < np; k++) {
7. // Compute the potential energy and forces.
8. //other code
9.    for ( j = 0; j < np; j++ ){
10.       if ( k != j ){
11.          //other code
12.      }}
13.   // Compute the kinetic energy
14.   for ( i = 0; i < nd; i++ ) {
15.      ke = ke + vel[i+k*nd] * vel[i+k*nd];
16.   }}}
17.   ke = ke * 0.5 * mass;
18.   *pot = pe;
19.   *kin = ke;
20.   return
21.   }
```

**Fig. 9** Snippet of OpenMP code generated by IPT—MD Simulation application

We demonstrate the usage of ICAT as a decision-support system by using it for porting the OpenMP version of the MD simulation application to KNL processors.

Before invoking ICAT, we compiled the OpenMP version of the MD simulation application with the -g flag. After invoking ICAT from the command-line, as shown in Fig. 10, we provide the path to the application executable and the arguments required to run it. We also select the advisor mode in which ICAT should run. Using this information, ICAT first profiles the application by running it in real-time

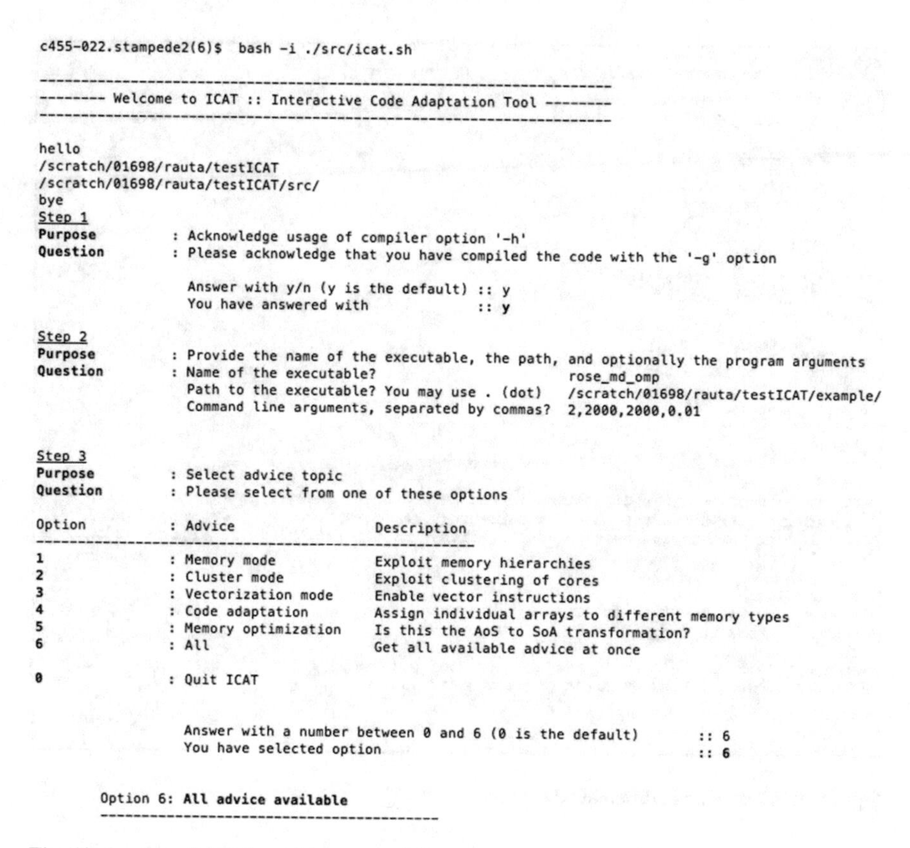
```
c455-022.stampede2(6)$  bash -i ./src/icat.sh

------------------------------------------------------------------
-------- Welcome to ICAT :: Interactive Code Adaptation Tool --------
------------------------------------------------------------------

hello
/scratch/01698/rauta/testICAT
/scratch/01698/rauta/testICAT/src/
bye
Step 1
Purpose         : Acknowledge usage of compiler option '-h'
Question        : Please acknowledge that you have compiled the code with the '-g' option

                  Answer with y/n (y is the default) :: y
                  You have answered with             :: y

Step 2
Purpose         : Provide the name of the executable, the path, and optionally the program arguments
Question        : Name of the executable?                  rose_md_omp
                  Path to the executable? You may use . (dot)   /scratch/01698/rauta/testICAT/example/
                  Command line arguments, separated by commas?  2,2000,2000,0.01


Step 3
Purpose         : Select advice topic
Question        : Please select from one of these options

Option          : Advice                 Description
-------------------------------------------------------
1               : Memory mode            Exploit memory hierarchies
2               : Cluster mode           Exploit clustering of cores
3               : Vectorization mode     Enable vector instructions
4               : Code adaptation        Assign individual arrays to different memory types
5               : Memory optimization    Is this the AoS to SoA transformation?
6               : All                    Get all available advice at once

0               : Quit ICAT

                  Answer with a number between 0 and 6 (0 is the default)      :: 6
                  You have selected option                                      :: 6

        Option 6: All advice available
        -----------------------------------------
```

**Fig. 10** Invoking ICAT from the command-line

and gathers the application's characteristics. It then generates a recommendation report regarding the appropriate memory mode for the application and instructions for compiling the code.

Using the memory mode report that it generated and the input regarding the programming model of the application, ICAT also generates a report with the recommendation for the cluster mode to use. Then, as shown in Fig. 11, ICAT informs that the entire MD simulation application fits in the MCDRAM. Hence, no code adaptation is required. However, if the user still wants to see how the code would be adapted to selectively use the MCDRAM, they may choose to do so by selecting the appropriate option while ICAT is running.

The reports generated by the memory mode advisor and the cluster mode advisor are shown in Figs. 12 and 13. For the OpenMP version of the MD simulation application, ICAT recommends running the application on the KNL node which has the MCDRAM configured in flat-mode if `numactl` is available. If `numactl` is not available, it recommends running the application on the KNL node which has the MCDRAM configured in cache-mode. For the cluster mode, ICAT recommends using the SNC-4 configuration. In the case of Stampede2 system,

```
Does your code use MPI programming model? (Enter 1 or 2.)
1. Yes
2. No
2
You chose 2
Profiling program...
Running perf command ...
Running the program again...
Report generated.
---------------------

Determining the clustering mode...
What is the programming model used in your application?
1. OpenMP
2. MPI
3. OpenMP + MPI
4. None of the above/serial
1
Report generated.
---------------------

_

Either the source code modification is not needed or the Memory Advisor report for
rose_md_omp does not exist in the subdirectory named reports. However, if you would
like to test how our source code modification script works, press 2, else press 3.
3

Please note:
If your code was modified by ICAT to take advantage of MCDRAM, then please compile it
with the -lmemkind flag.

You can run the code in the queue that is configured with MCDRAM in Flat mode or in
hybrid mode (e.g., Flat-Quadrant queue on Stampede).

_
```

**Fig. 11** ICAT running in different advisor modes

```
$ cat rose_md_omp_memory_advisor_report.txt
----- rose_md_omp Characteristics -----
Memory usage: 0.0216904 Cache Miss Rate: 0.726984

----- Recommendations -----

Application fits into HBM.

Mode to use: If numactl is available, use the Flat-Mode with all allocations to HBM.

If numactl is not available, then use the Cache-Mode. However, note that the cache misses
in the Cache-Mode are more expensive than reading data from DDR4 in Flat-Mode.

Memory Allocation: HBM
To execute the application in Flat-Mode: Use command < numactl --membind=1 ./run-app> if
it is serial, or < ibrun --membind=1 ./run-app > if it is parallel.

To execute the application in Cache-mode: Use the command that you normally use, that is,
< ./run-app > if it is serial or < ibrun ./run-app > if it is parallel.

In general, to determine the <NUMA_NODE> in the command < numactl --membind=NUMA_NODE > ,
run the command < numactl -H > and look for the node without any core

 ----- End ICAT report for rose_md_omp -----
```

**Fig. 12** Memory mode advisor report

```
$ cat rose_md_omp_clustering_advisor_report.txt
----- rose_md_omp Recommendations -----
 Clustering mode to use: SNC-4
Must pin threads using the following commands prior to executing program:
export OMP_NESTED=1
export OMP_NUM_THREADS=4,64
export OMP_PLACES=`numactl -H | grep cpus | awk '(NF>3) {for (i = 4; i <= NF; i++) printf
"%d,", $i}' | sed 's/.$//'`
export OMP_PROC_BIND=spread,close

numactl -m 4,5,6,7 ./run-app

 ----- End ICAT report for rose_md_omp -----
```

**Fig. 13** Cluster mode advisor report

```
#include <hbwmalloc.h>
#include <omp.h>
# include <stdlib.h>
//other code
...
int main(int argc,char *argv[]);
void compute(int np,int nd,double pos[],double vel[],double mass,double f[],double
*pot,double *kin);
//other code
...
int main(int argc,char *argv[]){
//other code
int checkHBMAvailability=hbw_check_available();
 if (checkHBMAvailability == 0){ acc = ((double *)(hbw_malloc(((nd * np) * sizeof(double
))))); } else{    acc = ((double *)(malloc(((nd * np) * sizeof(double ))))); }

//other code
...
if (checkHBMAvailability == 0){ hbw_free(acc); } else{   free(acc); }
//other code
...
}
```

**Fig. 14** Code modifications done using ICAT

the aforementioned recommendations imply that the OpenMP version of the MD simulation application should be run on the KNL node in the "Flat-SNC4" queue.

As can be noticed from Fig. 11, ICAT recommended against modifying the OpenMP version of the MD simulation application to use the HBWMALLOC interface. However, if a user still wishes to modify the application code to use the HBWMALLOC interface, they may do so using ICAT. A snippet of the modified version of the MD simulation application produced using ICAT is shown in Fig. 14. The modifications made by ICAT include: inserting code for including the `hbwmalloc.h` file, checking the availability of MCDRAM in the underlying architecture, replacing the call/s to the `malloc` function with the `hbw_malloc` function, and replacing the call/s to the `free` function with the `hbw_free` function call/s.

## 7 Conclusion

HPC system hardware and the programming models are constantly evolving. Sometimes the changes are big, but often the changes are incremental. Even if most users are not aiming for peak performance, they need to spend some effort in modernizing their applications to keep up with the major developments. Not modernizing their code base to keep up with the technology is not viable because with inefficient code, the researchers will neither be competitive (a) in the scientific arena to handle larger problem sizes and calculations than what they are doing now, nor (b) when HPC resources are allocated at open-science data centers [13]. Often, users attempt to strike a balance between effort and reward and they cannot afford to explore all possible avenues for code modernization at a given point in time. Therefore, high-level tools — like IPT and ICAT — that are geared towards assisting the users in code modernization and migration efforts on the latest HPC platforms are needed.

**Acknowledgements** We are very grateful to the National Science Foundation for grant # 1642396, ICERT REU program (National Science Foundation grant # 1359304), XSEDE (National Science Foundation grant # ACI-1053575), and TACC for providing resources required for this project. We are grateful to our students (Madhav Gupta, Trung Nguyen Ba, Alex Suryapranata, Julio Olaya, Tiffany Connors, Ejenio Capetillo, and Shweta Gulati) for their contributions to the IPT, ITALC, and ICAT codebase. We are also grateful to Dr. Purushotham Bangalore for providing guidance and code templates for developing FraSPA, which was the precursor of IPT.

## References

1. Arora, R., Olaya, J., Gupta, M.: A tool for interactive parallelization. In: Proceedings of the 2014 Annual Conference on Extreme Science and Engineering Discovery Environment (XSEDE'14), Article 51, p. 8. ACM, New York (2014). http://dx.doi.org/10.1145/2616498.2616558
2. Arora, R., Koesterke, L.: Interactive code adaptation tool for modernizing applications for Intel Knights Landing processors. In: Proceedings of the 2017 Conference on the Practice & Experience in Advanced Research Computing (PEARC17). ACM, New York (2017) http://dx.doi.org/10.1145/3093338.3093352
3. MPI: A Message Passing Interface Standard. Message Passing Interface Forum (2015). http://mpi-forum.org/docs/mpi-3.1/mpi31-report.pdf. Cited 16 June 2017
4. OpenMP Application Programming Interface (2015). http://www.openmp.org/wp-content/uploads/openmp-4.5.pdf. Cited 16 June 2017
5. CUDA Toolkit Documentation (2017). http://docs.nvidia.com/cuda/#axzz4lGuUKK2x. Cited 16 June 2017
6. ROSE User Manual (2017). http://rosecompiler.org/ROSE_UserManual/ROSE-UserManual.pdf. Cited 16 June 2017
7. Sodani, A.: Knights landing (KNL): 2nd generation Intel® Xeon Phi processor. In: 2015 IEEE Hot Chips 27 Symposium (HCS) (2015). doi:10.1109/HOTCHIPS.2015.7477467
8. Intel Corporation HBWMALLOC (2015). https://www.mankier.com/3/hbwmalloc. Cited 16 June 2017

9. perf: Linux Profiling with Performance Counters. https://perf.wiki.kernel.org/index.php/Main_Page. Cited 16 June 2017
10. Vtune Performance Profiler (2017). https://software.intel.com/en-us/intel-vtune-amplifier-xe. Cited 16 June 2017
11. Rapaport, D.: An introduction to interactive molecular-dynamics simulation. Comput. Phys. **11**(4), 337–347 (1997)
12. Molecular Dynamics Code. http://people.sc.fsu.edu/~jburkardt/c_src/md/md.c. Cited 16 June 2017
13. XSEDE Allocations Overview. https://www.xsede.org/allocations. Cited 16 June 2017

# Vectorization of High-Order DG in Ateles for the NEC SX-ACE

**Harald Klimach, Jiaxing Qi, Stephan Walter, and Sabine Roller**

**Abstract** In this chapter, we investigate the possibilities of deploying a high-order, modal, discontinuous Galerkin scheme on the SX-ACE. Our implementation Ateles is written in modern Fortran and requires the new sxf03 compiler from NEC. It is based on an unstructured mesh representation that necessitates indirect addressing, but allows for a large flexibility in the representation of geometries. However, the degrees of freedom within the elements are stored in a rigid, structured array. For sufficiently high-order approximations these data structures within the elements can be exploited for vectorization.

## 1 Introduction

Memory has become the limiting factor in most computing systems for most computations. Both, processing speeds and memory access has exponentially increased during the development of computing technology, albeit with different paces. This development led to a gap between the memory and processing capabilities in modern devices [1]. The important factor describing this relation for numerical applications based on floating point numbers is the Byte to FLOP (floating point operation) ratio. It can be used to judge the suitability of a system for a given algorithm. Because on the one hand, the computing system is capable of providing a fixed Byte to FLOP ratio, while on the other hand, the algorithm requires a certain amount of data to be moved for each operation.

Besides the speed of the memory and the number of required transactions, another important factor is the size of the memory. In comparison to the processing speed the amount of available main memory in high-performance computing systems did not increase much over the last decade. The largest amount of total memory provided by a *Top500* system is 1.5 PetaBytes on the *Sequoia* IBM

H. Klimach (✉) • J. Qi • S. Roller
University of Siegen, Adolf-Reichwein Str. 2, 57076 Siegen, Germany
e-mail: harald.klimach@uni-siegen.de; jiaxing.qi@uni-siegen.de; sabine.roller@uni-siegen.de

S. Walter
Höchstleistungsrechenzentrum Stuttgart, Nobelstr. 19, 70569 Stuttgart, Germany
e-mail: walter@hlrs.de


M.M. Resch et al. (eds.), *Sustained Simulation Performance 2017*,
DOI 10.1007/978-3-319-66896-3_5

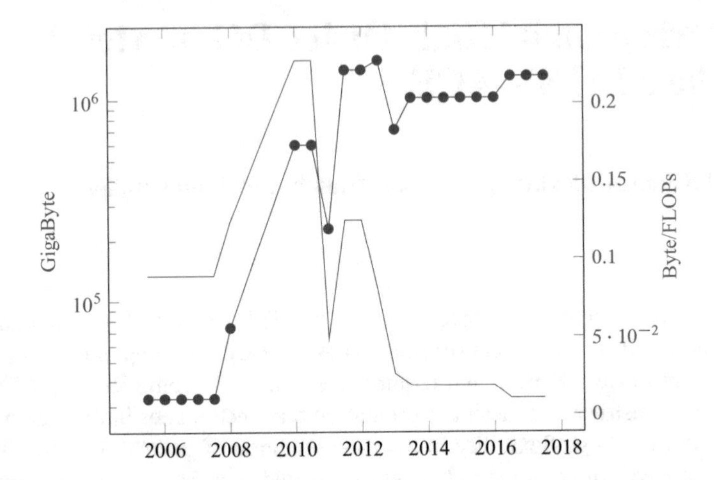


**Fig. 1** Development of available memory in the top system of the *Top500* list over time. In *blue* the total available memory of the fastest system from the list at that point in time is shown. In *red* the ratio between that memory and the computing power of the system is in terms of floating point operations per second is indicated

BlueGene/Q installation at the Lawrence Livermore National Laboratory. This was the fastest system in terms of floating point operations per second in 2012. In June 2017 it was ranked fifth in the *Top500*.

Figure 1 illustrates the development of the fastest system in the *Top500* lists with respect to the available memory over time. The blue trend with the dot markers indicates the available overall memory in the fastest system on a logarithmic scale. To put this in context to the available computing power of the system, the red line indicates the ratio between available main memory and the number of floating point operations per second. We can see that the overall amount of memory, which allows us to solve larger or better resolved problems only grows slowly, and even as it grows it does not keep pace with the computing speed of the systems. Thus, we can observe that memory is a precious resource in modern computing systems, both in terms of speed and of size. Furthermore, we even expect increasing importance of the memory in the foreseeable future as with the current growth rates the gap between processing and memory speeds will continue to grow.

Table 1 provides an overview to the memory properties of contemporary HPC architectures in relation to their floating point operation speed. The first column indicates the system, the second the memory bandwidth in Bytes per second divided by the floating point operations per second. In the last column the available amount of main memory, again divided by the number of floating point operations per second. This is the measure for which the development over time for the fastest system in the *Top500* is given in Fig. 1. The issues we face with the big amounts of data produced in large scale simulations only get worse when we actually want to store results. Storage devices are even slower than main memory and when considering time dependent data, we often need to store several snapshots of the overall main memory used by the simulation.

**Table 1** Contemporary HPC-systems with respect to their memory size and speed compared to their floating point operations per second

| System | Per FLOPS memory- | |
|---|---|---|
| | Bandwidth | Size |
| NEC SX-ACE | 1.000 | 0.250 |
| K Computer (SPARC64 VIIIfx) | 0.500 | 0.032 |
| Sequoia (IBM BlueGene Q) | 0.208 | 0.078 |
| Sunway TaihuLight (SW26010) | 0.178 | 0.010 |
| nVidia Tesla P100 | 0.138 | 0.003 |
| Hazel Hen (Intel Xeon E5-2680 v3) | 0.141 | 0.133 |

We see that the memory is slow and small when compared to the computing power in terms of performed operations. Therefore, an important criterion for numerical schemes to be deployed on such modern large-scale computing systems is their ability to provide good approximations with as little amount of memory as possible.

The discontinuous Galerkin (DG) scheme is a promising numerical method that enables us to move into the desired direction of reduced memory consumption for solutions of partial differential equations. It employs a discretization of the simulation domain by a mesh, where the solution within each element of the mesh is approximated by a local function. A typical choice for the functions to use in this approximation are polynomial series. The usage of functions to represent the solution allows for high-order representations, as the scheme works for arbitrary numbers of terms in the deployed functions. High-order approximations have the advantage that they can approximate smooth functions with few degrees of freedom, due to the exponential convergence with increasing number of modes. Thus, the scheme requires only a minimal amount of data to represent the solution in the elements. Interaction between elements is realized by fluxes like in finite volume schemes. The discontinuous Galerkin scheme, thereby, offers a combination of aspects from the finite volume method and spectral discretizations. It provides to some extend the efficiency of spectral methods and at the same time some of the flexibility offered by finite volume methods.

From the numerical side the discontinuous Galerkin scheme appears to provide suitable characteristics to address the growing imbalance between memory and processing power of modern computing systems. On the side of computing architectures, the NEC SX-ACE is a vector system that offers some nice capabilities for numerical schemes with a focus on good memory performance. It offers a high Byte per FLOP ratio of 1 (256 GFLOP and 256 GB per second) with access to 16 GB of main memory per core with this speed when using all 4 cores of the processor. If this Byte to FLOP ratio is insufficient for an algorithm, it can be increased up to 4 Bytes per FLOP by employing fewer cores of the processor in the computation. As can be seen in Table 1 this is at the high end of this ratio for contemporary HPC architectures.

Because of these properties, we believe the discontinuous Galerkin method and the *NEC SX-ACE* architecture are a good match for large-scale simulations. The one provides an option to reduce the memory usage and the other attempts to provide a high data rate to allow a wider set of applications to achieve a high sustained performance. However, an obstacle we face in typical applications is the need for a great deal of flexibility and dynamic behavior during the runtime of the simulation. This often does not fit too well with the more rigid requirements for efficient computations on vector systems. Here we want to lay out, how the discontinuous Galerkin scheme with a sufficiently high order may be used to combine the flexibility required by the application with the vectorized computation on the NEC SX-ACE. This possibility is opened by the two levels of computation present in the discontinuous Galerkin scheme, where we can find high flexibility on the level of the mesh, but a highly structured and rigid layout within the elements. We believe, that it is a feasible option to use vectorization within elements of high-order discontinuous Galerkin schemes, while maintaining the large flexibility, offered by the method on the mesh level. Such a strategy opens the possibility to combine dynamic and adaptive simulations with the requirements of vectorized computing, which is increasingly important also on other architectures than the NEC SX-ACE.

In the following we briefly introduce the high-order discontinuous Galerkin scheme implemented in our solver *Ateles*. Then we go on with the presentation of the vectorization approach of the scheme on the NEC SX-ACE in Sect. 3 and conclude this chapter with some measurements and observations in Sect. 4.

## 2 High-Order Discontinuous Galerkin in Ateles

The discontinuous Galerkin method is especially well suited for conservation laws of the form:

$$\frac{\partial u}{\partial t} + \nabla f(u) = g \tag{1}$$

To find a solution to (1), the overall domain to be investigated is split into finite elements $\Omega_i$ and the solution is approximated by a function $u_h$ within each of these elements. The equation is then multiplied with test functions $\phi$ to create a system that can be solved and after integration by parts we obtain:

$$\frac{\partial}{\partial t}\int_{\Omega_i} u_h \phi \mathrm{d}V - \int_{\Omega_i} f(u_h)\nabla\phi \mathrm{d}V + \int_{\partial\Omega_i} f^* \phi \mathrm{d}S = \int_{\Omega_i} g\phi \mathrm{d}V \tag{2}$$

From the integration by parts we get the surface integral where the new term $f^*$ is introduced. This is a numerical flux that ties together adjacent elements as it requires the state from both sides of the surface. In a numerical discretization the employed function spaces for the solution and the test functions need to be finite and Eq. (2) then provides an algebraic system in space, where the products of the functions can be written in matrices. Especially we get the mass matrix:

$$M = \int \psi\phi \mathrm{d}V \tag{3}$$

and the stiffness matrix:

$$S = \int \psi\nabla\phi \mathrm{d}V \tag{4}$$

## 2.1 The Modal Basis

*Ateles* implements the discontinuous Galerkin scheme with Legendre polynomials as a basis to represent the solution $u$ in cubical elements. The Legendre polynomials can be defined recursively by:

$$\begin{aligned} L_0(x) &= 1, \quad L_1(x) = x \\ L_k(x) &= \frac{2k-1}{k} \cdot x \cdot L_{k-1}(x) - \frac{k-1}{k} L_{k-2}(x) \end{aligned} \tag{5}$$

They are defined on the reference interval $[-1, 1]$ and have some favorable properties. Most importantly they build an orthogonal basis with respect to the inner product with a weight of 1 over this interval. Another nice property is that all Legendre polynomials except for $L_0(x)$ are integral mean-free. Our solution within the elements of the discontinuous Galerkin scheme are obtained in the form of a series of Legendre polynomials:

$$u(x) = \sum_{k=0}^{m} c_k L_k(x) \tag{6}$$

Here, the coefficients $c_k$ are the (Legendre) modes that describe the actual shape of the solution. The maximal polynomial degree in this series is denoted by $m$. Its choice determines the spatial convergence order of the scheme and the degrees of freedom (modes) required to represent the solution in the element $(m + 1)$. To represent the solution in three-dimensional space, we build a tensor product of the one-dimensional polynomials. By introducing the multi-index $\alpha = (i, j, k)$, we can denote the three-dimensional solution by:

$$u(x, y, z) = \sum_{\alpha=(0,0,0)}^{(m,m,m)} c_\alpha L_i(x) L_j(y) L_k(z) \tag{7}$$

With this definition for the solution in $d$ dimensions, we get $(m + 1)^d$ degrees of freedom. The layout of this data is highly structured, as we need a simple array with $(m+1) \times (m+1) \times (m+1)$ entries to store the $c_\alpha$ in three dimensions for example.

The orthogonality of the Legendre polynomials enables a fast computation of the mass matrix and its inverse, and their recursive nature enables a fast application of the stiffness matrix.

### 2.2 The Mesh Structure

The local discretization by polynomial series as described above is done locally in elements that are then combined in a mesh to cover the complete computational domain. *Ateles* employs an octree topology to construct this mesh of cubical elements with an unstructured layout. The unstructured organization requires an explicit description of elements to be considered but allows for a greater flexibility in describing arbitrary geometrical setups. By relying on an octree structure, large parts of the topological information is implicitly known and does not have to be explicitly stored or referred to. This is especially of an advantage for distributed parallel computations, as most neighbor information can be computed locally with a minimal amount of data exchange. With the choice of cubical elements, we can employ an efficient dimension by dimension approach and avoid the need for complex transformations. Boundaries are then implemented by penalizing terms inside the elements, very similar to approaches found in spectral discretizations. These allow for the approximation of the geometry with the same order as the one used for the representation of the scheme.

## 3 Vectorization on the NEC SX-ACE

Vector instructions mean that we perform the same instruction to many data concurrently. This single instruction, multiple data (SIMD) concept is becoming more and more important also on traditional scalar systems, as can be seen in the increasing register lengths of the AVX instructions in Intels x86 architecture. The NEC SX-ACE as a traditional vector computing system offers long vector data registers that hold 256 double precision real numbers and can perform one instruction on all of them simultaneously. From the algorithmic point we need long loops with independent iterations to utilize this mechanism.

In simulations that involve meshes, we usually need to perform the same operations for each mesh element, and we have many mesh elements for detailed simulations. Thus, an obvious choice for vectorization is here the loop over elements of the mesh. However, for high-order schemes this is not so straight forward anymore. For one, there are fewer elements used in the discretization, and maybe even more important, the computation for each element gets more involved. The greatest problem for an efficient vectorization over the elements, however, is the desire for flexibility on the level of the mesh. As described above, we use an unstructured mesh description to enable an efficient approximation of arbitrary geometries. This introduces an indirection, which is in turn hurting the performance, as the vector data needs to be gathered and scattered when moved between memory and registers. Even more flexibility is required on the mesh level, when we allow hp-adaptivity, that is dynamic mesh adaptation to the solution and a variation in the polynomial degree from element to element. These features are desirable, because they minimize the computational effort in terms of memory and operations.

With this large degree of flexibility and unstructured data access across the elements of the mesh, a vectorized computation appears hard to achieve. Instead we look here into the vectorization within elements. As described in Sect. 2 the data within elements is highly structured and the operations we need to perform on it also nicely fits into SIMD schemes for a large part. One of the main computational tasks is the application of the stiffness- and mass-matrices. Such matrix-vector multiplications can be perfectly performed in vector operations. Other numerical tasks within the elements often follow a similar scheme and require the application of one operation to all degrees of freedom. The main limitation we face with an approach of vectorization within elements is the limited vector length. However, the vector length grows with the polynomial degree, opening the possibility to fully exploit even long vector registers, if the polynomial degree is only sufficiently high.

Most operations in *Ateles* need to be done on the polynomials in one direction, leaving the other dimensions open for concurrent execution. Thus, when the solution in a three-dimensional element is approximated by a maximal polynomial of degree $m$, there are $(m + 1)^3$ degrees of freedom in total, and in most operations $(m + 1)^2$ independent computations with the same instruction need to be performed. With this quadratic growth over the polynomial degree, $m = 15$ is already sufficient to fill the vector data registers with a length of 256 for the most important parts of the implementation. For the high-order discretization in *Ateles* we aim for polynomial degrees greater than 10, and for linear equations even for polynomial degrees in the range of 100. With this range of scheme orders, a vectorization within elements appears suitable and meaningful, even for such long vectors as found in the NEC SX-ACE.

The use of polynomials of high degree to represent the solution, thereby enables us to combine the flexibility of mesh adaptivity and unstructured meshes with efficient vector computations.

## *3.1 Porting of Ateles*

*Ateles* is implemented in modern *Fortran* and utilizes some features from the *Fortran 2003* standard. Unfortunately, the existing Fortran compiler from older SX systems did not provide all the required features and was unable to compile *Ateles*. But NEC has implemented a new compiler for the SX-ACE, which supports the complete *Fortran 2003* standard. This new compiler *sxf03* was able to compile *Ateles* and create a working executable for the SX-ACE, with surprisingly little effort. Yet, as this is a new compiler, not all optimizations from the old compilers where initially available and after the first porting, we ran into a vectorization issue with one of the loops, that was nicely vectorized by the old compiler, but not by the new *sxf03*. Because compiled files from the old and the new compiler could not be combined, the work on further optimization stalled at that point. A little more on these first porting issues can be found in [2] from last year, where also some more explanations on the porting of the APES suite in general are provided. After this issue was fixed in the compiler by NEC, we were now able to further look into the vectorization of *Ateles* and how the vectorization strategy within elements works out. In the following we report on the progress of this effort.

# 4 Measurements and Observations

To compile *Ateles* for the NEC SX-ACE in this report, we make use of the *sxf03* compiler in version "*Rev.050 2017/01/06*". As explained in Sect. 3, we are concerned with the operations within elements, and most of those resemble matrix-vector operations or are quite similar to them. One major distinction can be drawn depending on the kind of equation system that we need to solve. For linear equations we can perform all numerics in modal space, directly using the terms from the polynomial series, as introduced in Eq. (6). When dealing with nonlinear equations this is not so easily possible anymore. Instead, we transform the representation into physical space to obtain values at specific points, perform the nonlinear operation in each point and then transform the new values back into modal representation again. These transformations need to be done additionally and are quite expensive. The performance characteristics of the two cases are accordingly largely different in these two cases.

## *4.1 Linear Equations*

Let us first look at linear equations, as their building blocks are also relevant for the nonlinear equations. As a representative for linear equations we look into the Maxwell equations for electrodynamics. We use a simple case without boundary conditions and polynomials of degree 11. All computations are done on a single core of the SX-ACE. In our first setup we used 64 elements, and found a really poor performance of only 26 MFLOPS in the most expensive routine according to the *ftrace* analysis. The crucial loop of that routine is shown in Listing 1 and we would expect this to nicely vectorize the inner, collapsed loop. Indeed, the extremely poor performance was due to the number of elements, as this is the first index here, and we end up with a strided access, according to the number of elements. When

**Listing 1** Main loop of the volume to face projection

```
do iAnsZ=1,m+1,2
  !collapsed loop
  do iVEF=1,6*nElems*(m+1)**2
    ! indices actually computed from iVEF
    facestate(iElem,facepos,iVar,side) &
      & = facestate(iElem,facepos,iVar,side) &
      & + volstate(ielem,pos,iVar)
  end do
end do
```

**Table 2** Excerpt from the tracing of *Ateles* for Maxwell equations and a discretization with polynomials of degree 11

| Procedure | % | MFLOPS | V.OP % | V.LEN | Bank Conf. CPU | Bank Conf. Net | ADB % |
|---|---|---|---|---|---|---|---|
| VolToFace | 28.7 | 929.5 | 99.12 | 204.6 | 0.137 | 0.531 | 86.07 |
| PrjFlux2 | 10.3 | 1475.0 | 99.53 | 83.6 | 0.802 | 1.742 | 79.04 |
| PrjFlux1 | 10.3 | 1479.4 | 99.53 | 83.2 | 0.685 | 1.294 | 71.65 |
| PrjFlux3 | 10.2 | 1490.8 | 99.57 | 83.4 | 0.381 | 1.752 | 76.73 |
| MaxFlux | 9.9 | 513.3 | 79.88 | 38.7 | 0.057 | 0.321 | 63.90 |
| MassMat | 9.3 | 1906.7 | 87.69 | 63.0 | 0.503 | 18.491 | 45.46 |
| PhysFlux | 5.5 | 0.2 | 94.54 | 241.3 | 0.941 | 7.973 | 0.00 |

The first column states the measured routine, the second the running time percentage of the routine, the third the observed MFLOPS, the fourth is the vector operation ratio as a percentage (time spent on vector instructions to the time spent in total on that routine) and the fifth column provides the average vector length used in the vector instructions. The next two columns (6 and 7) provide the time spent on conflicts when accessing memory banks. In the eighth and last column, the ADB hit rate is given. Shown are the main procedures contributing to the overall compute time

changing to the element count to 63, the performance indeed increases from 26 to more than 900 MFLOPS. It appears that strides at multiples of 64 result in extremely bad performance, due to conflicts in the memory bank accesses.

Table 2 shows the most important routines for a run with polynomials of degree 11 and 63 elements. The main routines that contribute more than 84% to the overall compute time are the projection of the polynomials in the volume to the faces of the elements (*VolToFace*), the projection of the fluxes onto the testfunctions (*PrjFlux1*, *PrjFlux2* and *PrjFlux3*), the actual computation of the Maxwell flux (*MaxFlux*), multiplication with the inverse of the mass matrix (*MassMat*) and computation of the physical flux for the Maxwell equations (*PhysFlux*). Here, the projection of the flux onto testfunctions is actually the same operation that needs to be performed, albeit in three different directions and there is an individual implementation for each direction. Their only distinction is a different striding in the access to the three dimensional data.

As can be seen in Table 2, the volume to face projection (*VolToFace*) and the projection of the physical flux on the testfunctions are the main consumers of the computing time in this run with polynomials of degree 11. Both contribute about 30% to the overall running time. It also can be seen that already this run without tuning, provides relatively good vectorization properties with vectorization rates above 99%. Nevertheless, the computational efficiency is not quite high and we see for the *VolToFace* routine less than one GFLOPS. But this may also be due to the relatively low computational density in this operation. What needs to be done is just the summation of the degrees of freedom in one space direction. This single addition for each real number does not allow us to fully exploit the functional units of the processor.

One improvement that can be done in this routine is the simultaneous computation of the left and right faces in the given direction of the element. This improves the computational density as the volume data only needs to be loaded once for both sides, and we obtain between 1262 and 1462 MFLOPS (depending on the striding for the different directions). After this change, the projection of the physical fluxes becomes the most time consuming part. When we double the degrees of freedom and use polynomials of degree 23, this changes again and the multiplication with the inverse of the mass-matrix becomes the most important routine. Computing the multiplication with the inverse of the mass-matrix makes use of a short recursion, as with the recursive definition of the polynomial basis, already computed values can be reused. While this is computational efficient in terms of saving operations, it makes it harder to achieve good vectorization and a high sustained performance. Yet, for high orders and when avoiding bad striding we are capable to achieve already reasonable performance and before looking into this common part in more detail we now looked into other equations.

Unsurprisingly the acoustic and linearized Euler equations showed a very similar behavior. However, we found an excessive use of flux functions there to be an issue. This is already a little bit visible in Table 2 for the *PhysFlux* routine. The problem with that routine is that it is very small and used to compute the flux for just a single mode. Similarly this was found for the other linear equations, but there it the flux computation consumed a larger fraction of the overall compute time, and the effect was more pronounced. The remedy is fairly simple, though, as the loop over the modes can be pulled into this routine quite easily.

Another linear equation we have implemented in *Ateles* are locally linearized Euler equations. These use a linearization within elements, but nonlinear fluxes on the element faces for the exchange between elements. With those we ran again into the striding issue with the number of elements. Further investigation revealed that this striding indeed is the most important factor inhibiting better sustained performance on the SX-ACE. Our findings for the linear equations reinforced the idea that we need to do the vectorization within the elements, and we now need to change the data structure to reflect this, as the element index is often the fastest running index in our arrays. This will be a larger effort and instead we now turn to the nonlinear equations and have a brief look at the performance of the inviscid Euler equations for compressible flows.

## 4.2 Nonlinear Equations

To investigate nonlinear terms, we look here into the Euler equations for compressible fluid flows. As mentioned, we need to perform polynomial transformations between modal and nodal space in this case. There are several methods for this task implemented in *Ateles*, see [3] for more details and a comparison of the methods. For now we will only consider the $L_2$ projection of the Legendre modes to their nodes (L2P transformation in *Ateles*). This method is the most straight-forward one and can simply be written as a matrix-vector multiplication.

We look at the Euler equations for inviscid, compressible flows here. The original code showed no performance, due to the fluxes being called for each integration point and ending up to be the most time-consuming parts with only little to none FLOPS achieved. By pulling the loops over the points into the flux computation, this can be avoided and the contribution of these routines to the total computational time becomes negligible for now. Instead the *VolToFace* and the L2P are the most important contributors.

As the $L_2$ projection basically is a matrix-vector product, we also see a relatively high performance for this routine of more than 13 GFLOPS. However, we actually need to perform many of these matrix-vector products and it can be interpreted as a matrix-matrix product. If we rewrite our code such that the compiler recognizes this construct, it replaces it with a highly optimized implementation for the architecture and we gain close to 48 GFLOPS or 75% sustained performance for this operation when using polynomials of degree 29. To allow the compiler to recognize the construct, a two-dimensional array has to be used, which was previously not the case, as a collapsed index was used.

Table 3 shows the most relevant routines for Euler equations with a 30th order spatial discretization. The most efficient routine, the $L_2$ projection with the recognized matrix-matrix operation is also the most time consuming one, leading to a relatively good overall sustained performance. Further we recognize the volume to face projection, that was also relevant for the linear equations, but in its optimization has been split into three routines, one for each direction. Also the projection of the fluxes to the testfunctions are again contributing visibly to the overall compute time. The only other routine with more than 5% in the overall compute time is the *FromOver* routine, which implements the copying of the modal state to an oversampled space.

Thus, we see that further optimizations of the general routines that are important in the linear equations also will be beneficial for the nonlinear equations. We expect to achieve the next larger performance increment with the change of the index ordering for our state arrays.

Finally, we want to compare the serial performance of the current implementation status for varying orders to the observed performance on a scalar system. The scalar system we compare against is the Cray XC 40 *Hornet* at *HLRS* that is equipped with Intel Xeon E5-2680 v3 processors. To allow the comparison of runs on different machines and between different orders we use thousand degree of freedom updates

**Table 3** Excerpt from the tracing of *Ateles* for Euler equations and a discretization with polynomials of degree 29

| Procedure | % | MFLOPS | V.OP % | V.LEN | Bank Conf. | | ADB % |
|---|---|---|---|---|---|---|---|
| | | | | | CPU | Net | |
| L2project | 34.3 | 47916.4 | 99.21 | 174.5 | 1.680 | 6.543 | 75.38 |
| VolToFaceY | 6.9 | 1459.4 | 99.89 | 256.0 | 0.001 | 0.000 | 4.02 |
| FromOver | 6.5 | 2841.2 | 99.86 | 254.7 | 0.013 | 0.096 | 3.80 |
| PrjFlux3 | 6.5 | 2544.4 | 99.90 | 186.9 | 0.781 | 1.237 | 68.88 |
| VolToFaceX | 6.1 | 1312.0 | 99.87 | 256.0 | 0.001 | 0.000 | 1.51 |
| VolToFaceZ | 6.1 | 1316.1 | 99.88 | 256.0 | 0.012 | 0.000 | 2.44 |
| PrjFlux2 | 5.6 | 2943.7 | 99.89 | 187.0 | 0.511 | 0.504 | 70.55 |
| PrjFlux1 | 5.5 | 2992.5 | 99.89 | 186.8 | 0.470 | 0.366 | 26.54 |

The first column states the measured routine, the second the running time percentage of the routine, the third the observed MFLOPS, the fourth is the vector operation ratio as a percentage (time spent on vector instructions to the time spent in total on that routine) and the fifth column provides the average vector length used in the vector instructions. The next two columns (6 and 7) provide the time spent on conflicts when accessing memory banks. In the eighth and last column, the ADB hit rate is given. Shown are the main procedures contributing to the overall compute time

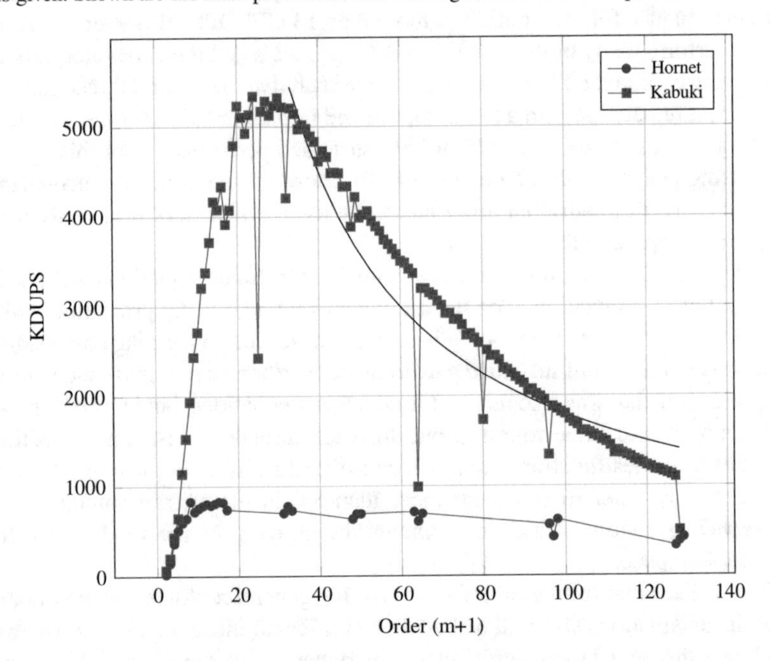


**Fig. 2** Performance for the Euler 3D equation, 1 process, 20 million degrees of freedom in total

per second (KDUPS). The performance for the Euler equation and a total of 20 million degrees of freedom is shown in Fig. 2 over varying polynomial degrees for a fixed overall problem size of a total of 20 million degrees of freedom. Note that the computational effort per degree of freedom grows with order of the scheme for nonlinear equations, and we expect a degradation of the degree of freedom update rate. This expected degradation is indicated by the black continuous line without marks. It does not imply a decrease in computational efficiency but rather the opposite, as we need to perform ever more operations without an increase in the data rate. As can be seen, a good performance is achieved when polynomials of degree 20 or higher are used for the nonlinear equations. We also notice several breakdowns of performance at some even order schemes. Most prominently for schemes of order 64, where we find the bad striding access described above, this time due to strided access over the modes of the polynomials within elements.

Of course there is not much of an benefit from the vectorization for scheme orders below 10 with a vectorization over within the elements, but already for 10 order schemes the vector computing capabilities can be used quite visibly. For very high orders beyond 100, the advantage seems to diminish again somewhat, but this seems to be more a point of the scalar system gaining efficiency due to the reduced bandwidth requirements in this range.

## 5 Summary and Outlook

High-order schemes are an attractive tool from the computational point of view, due to there reduced memory requirements. We presented a concept for the vectorization of the high-order discontinuous Galerkin scheme in *Ateles* with a focus on the structured data within elements to represent the three-dimensional polynomials. This approach enables a flexible computation on the mesh side with adaptivity and unstructured meshes, while at the same time allows for vectorized computations on highly structured data within the elements. Even with the long vectors of the NEC SX-ACE this concept works quite nicely already for relatively low order schemes with polynomials of degree 10 and higher.

Our current implementation is not yet tuned a lot for the vector system, and there are some legacy parts that need to changed, like the ordering of indices in the state representation. Nevertheless, the performance achieved on the SX-ACE already provides a quite good basis for further improvements. A particular positive surprise was the compiler optimization with the detected matrix-matrix multiplication construct and its optimized replacement by the compiler. Though, the nonlinear and linear equations have different routines that contribute to the overall computational effort, there is still a large overlap, and most further improvements are expected to affect all supported equations.

**Acknowledgements** We would like to thank Holger Berger from NEC for his kind support, the Tohoku University and HLRS for the opportunity to use their NEC SX-ACE installation.

# Vectorization of Cellular Automaton-Based Labeling of 3-D Binary Lattices

**Peter Zinterhof**

**Abstract** Labeling connected components in binary lattices is a basic function in image processing with applications in a range of fields, such as robotic vision, machine learning, and even computational fluid dynamics (CFD, percolation theory). While standard algorithms often employ recursive designs that seem ill-suited for parallel execution as well as being prone to excessive memory consumption and even stack-overflows, the described new algorithm is based on a cellular automaton (CA) that is immune against these drawbacks. Furthermore, being an inherently parallel system in itself, the CA also promises speedup and scalability on vector supercomputers as well as on current accelerators, such as GPGPU and Xeon PHI.

## 1 Introduction

Labeling connected components in binary lattices is a basic function in image processing with applications in a range of fields, such as robotic vision, machine learning, and even computational fluid dynamics (CFD, percolation theory). While standard algorithms often employ recursive designs that seem ill-suited for parallel execution as well as being prone to excessive memory consumption and even stack-overflows, the described new algorithm is based on a cellular automaton (CA) that is immune against these drawbacks. Furthermore, being an inherently parallel system in itself, the CA also promises speedup and scalability on vector supercomputers as well as on current accelerators, such as GPGPU and Xeon PHI.

The discussed algorithm for finding connected components within 3-dimensional lattices is based on a Cellular Automaton (CA) [1] which is a classic and well-studied tool in computer science.

In general, Cellular Automata operate on a set of cells (e.g. pixels or data items) which are manipulated according to a set of transition rules, which can be applied to all cells sequentially or in parallel. The manipulation is repeated iteratively until

P. Zinterhof (✉)
Department of Computer Science, University of Salzburg, Jakob-Haringer-Str. 5, 5020 Salzburg, Austria
e-mail: peter@zinterhof.com


M.M. Resch et al. (eds.), *Sustained Simulation Performance 2017*,
DOI 10.1007/978-3-319-66896-3_6

**Table 1** Execution times (ms) of dense CA of varying dimensions

| CA dim | GTX680 | NEC ACE-SX | Tesla P100 | Xeon E1620 | Xeon Phi 5110 |
|---|---|---|---|---|---|
| 128 | 0.25 | 0.57 | 0.26 | 11.37 | n/a |
| 256 | 1.86 | 3.71 | 0.69 | 68.85 | n/a |
| 384 | 6.53 | 13.37 | 5.8 | 189.88 | n/a |
| 512 | 15.55 | 31.81 | 5.8 | 399.86 | n/a |
| 704 | 49.55 | 79.03 | 13.98 | 772.18 | 42.44 |

some stopping criterion has been reached, for instance an equilibrium condition in which no further changes do occur or some runtime constraint has been met.

Among a series of convenient characteristics of Cellular Automata we want to emphasize their decent memory requirements which in most cases will be fixed during runtime and proportional to the number of cells while being agnostic to cell states. Also, updating cells in a CA is an operation that shows very high degrees of data-locality, which by itself can be regarded as an important prerequisite in the context of implementations for massively parallel and even distributed systems.

We consider these properties to be quite an advantage over recursive algorithms for finding connected components, which display patterns of memory consumption that are related both to the number and states of lattice cells. This makes CA-based computation of connected components an attractive choice for tightly memory restricted computer systems, in some cases probably even the only viable choice. Additionally, two very important advantages of the proposed algorithm can be named by the homogeneity of computational intensity within the lattice of cells, and the high regularity of memory access patterns during the iterations of the algorithm. Both specifics lend themselves well to high performance implementations on parallel systems.

As the set of transition rules can be applied to all cells of the lattice in parallel, the computational core of the algorithm is inherently parallel, too.

Our main contribution is given by the definition and discussion of vectorized and parallel implementations of the basic CA-algorithm on a variety of recent vector- and parallel compute architectures (Table 1).

## 2 Related Work

This work is based on the important paper by Stamatovic and Trobec [4] which introduces a new method of computing connected components in binary lattices by application of the well-known theory of cellular automata[1] in a new way. Compared to [6] we concentrate on the 3-D case instead of the 2-D case of input data.

[1] A comprehensive introduction to CA theory can be found in [2].

**Table 2** Typical memory requirements (GB) for Matlab function 'bwconncomp' and reported CA-based implementations

| CA dim | Matlab | Cellular automaton |
|---|---|---|
| 256 | 1.83 | 0.13 |
| 512 | 11 | 1.0 |
| 768 | 29 | 3.6 |

The considerably higher RAM-requirements for Matlab's 'bwconncomp' function may also lead to swapping on some systems and CA dimensions, which will not occur in the CA-based counterpart

Stamatovic and Trobec [4] also covers the 3-D case but puts more emphasis on the discussion of algorithmic details and the general proof-of-concept by displaying implementations in Matlab and NetLogo while this contribution is focused on various aspects of high-performance implementations on parallel hardware. The well-known Matlab software environment also offers some built-in function 'bwconncomp' which is capable of computing connected components within multidimensional arrays. Despite being convenient to use, Matlab's implementation falls short with regards to performance, memory requirements (Table 2) and potential use of accelerators, such as GPUs.

Other related work includes the class of stencil-based algorithms, which are not widely regarded as siblings of CA theory but the field of numerical analysis. To name just a few, stencil-based algorithms are applied in areas such as computational fluid dynamics (CFD), Lattice-Boltzmann computations, Jacobi solvers, Simulation of heat-transfer (e.g. convection) and image processing. Due to the importance of stencils in computer science[2] there have been many approaches to improve computational performance of the core algorithm by means of vectorization and parallelization [3]. Initially, these approaches were mostly based on optimization of some given algorithm on some specific target system. This exhibited limitations both to portability and usability, as forthcoming developments in parallel and distributed systems technology or additional requirements on the algorithmic level implied deep changes to the initially optimum code bases and implementation details.

Various forms of code generators and definition languages (for more information also see the interesting work on the pochoir stencil compiler [5]) for stencil computations have been described, which essentially introduce some kind of abstraction layer between actual compute hardware and the mathematical definition of stencils. Due to ever increasing complexity in compute hardware, namely increasing number of memory levels that operate at different speeds and latencies, and increasing numbers of cores per system, the generation of high performance code is now a task that seems to overwhelm not only most human software developers but also many standard code generators. To alleviate this rather undesirable situation, special auto-tuning frameworks have been proposed. These frameworks aim to sift through the

[2]The large-scale research project 'Exastencils' (http://www.exastencils.org/) is also to be mentioned in this context.

enormous number parameters found in the implementation of some stencil code and find optimal settings automatically without requiring much domain-specific expertise by the user.

Despite this intriguing corpus of related work, we found very little support for the kind of mathematical operations that the proposed CA-based algorithm is based on. Also, most work on stencil-based algorithms are based on regular and dense datasets, while our approach complements computation of dense datasets by some sparse formulation of the CA update routines. To the best of our knowledge, we will give the first report on the application of this CA-based algorithm in the context of sparse lattices on parallel hardware.

## 3 Algorithm

The basic algorithm for computing a 3-dimensional Cellular Automaton for finding connected components in a lattice follows [6]. The main algorithmic steps show great similarity to a stencil computation, in which floating point data is exchanged by integer data and the numerical summation of stencil pixels is replaced by the computation of the maximum value within the stencil pattern.

Also, we restrict the 'observed neighborhood' of each cell to the Von-Neumann neighborhood, which is defined as a center- or 'host'-cell and two directly adjacent neighbor-cells for each dimension. Following Trobec and Stamatovic [6] the lattice boundary cells are fixed during cell updates. By employing this fixed boundary condition the resulting code complexity can be reduced, which is an advantage on most of the projected target platforms that we will consider in the following section.

On entry, the binary lattice describes a distribution of 'background' and 'foreground' values only. The task of finding connected components is accomplished by a short initialization phase, along with the actual Cellular Automaton update phase, which by itself is an iterative process.

### *3.1 Initialization*

During initialization the binary lattice data will be transformed into an initial configuration or 'coloring' in which each foreground pixel will be given a unique index value or 'color' that will enable a clear distinction of pixels inside the lattice. For performance reasons we refrain from the initial coloring scheme described by Trobec and Stamatovic [6], which takes into account so-called 'corner pixels' for setting up the initial lattice values. Instead we choose a strictly monotonous series of cardinal values that are attached distinctively to any binary foreground pixel. Albeit this approach is inherently sequential (Algorithm 1), we found it of sufficient speed. Alternatively, initial coloring can also be achieved by choosing the positional data of each lattice point, which is given by the distinct tuple of $(x, y, z)$-values from which some distinct cardinal value can readily be derived in parallel execution mode.[3] It is essential to have boundary cells being initialized to 'background' states. This can be accomplished either in the initial binary lattice data or in initialization step.

## 3.2 Cellular Automaton Update

The update rule for the lattice (CA1) is applied to each cell that has neither been labeled as 'background' nor is a member of the boundary of the lattice. Each iteration of the Cellular Automaton takes CA1, the current state of the CA, and yields a new lattice CA2 in which the states of non-background and non-boundary cells have been updated. The update rule is given by the maximum function, applied to the current host cell C and its 6-neighborhood of surrounding cells. As each to be updated host cell C might also be a neighbor cell, this newly computed cell state must not be stored at the current lattice position but in a disjoint memory location in state array CA2, hence the transform $CA1- > CA2$. Algorithm 3 gives an outline of the update process in pseudo code.

**Algorithm 1** CAinitialize

1: **procedure** CA_INITIALIZE(input: binary_lattice, output: CA)
2:     *color* ← 0
3:     **for all** cells in binary_lattice **do**
4:         **if** cell **then**
5:             *color* ← *color + 1*
6:             *CA.cell.color* ← *color*
7:             *return*(*CA*)

1: **procedure** MAXIMUM_NEIGHBOR(input: cell, output: color)
2:     *color* ← *cell.color*
3:     *color* ← *max (color, cell.up)*
4:     *color* ← *max (color, cell.down)*
5:     *color* ← *max (color, cell.left)*
6:     *color* ← *max (color, cell.right)*
7:     *color* ← *max (color, cell.before)*
8:     *color* ← *max (color, cell.after)*

[3] Let's consider a cubic lattice of dimension N. For any lattice element at some position $(x, y, z)$ the unique positional information can be used to derive an initial coloring $Color = (((z * N) + y) * N + x$.

**Algorithm 2** Termination check

1: **procedure** TERMINATION(input: CA1, input: CA2, output: bool)
2:     **for all** cells in CA1 **do**
3:         **if** (Ca1.cell.color NOT CA2.cell.color) **then return** *False*
    **return** *True*

**Algorithm 3** Cellular automaton update

1: **procedure** CA_UPDATE(input: CA1, output: CA2)
2:     **for all** cells in CA1 **do**
3:         **if** cell NOT (background OR boundary) **then**
4:             *CA2.cell.color* ← *maximum_neighbor (CA1.cell)*

### 3.2.1 Maximum Operator

Despite being a very basic operation, computing the maximum pixel values of the surrounding neighborhood of each cell constitutes the main part of the Cellular Automaton which usually will take most of the total runtime of the proposed algorithm on any given hardware platform. Hence, we aim to support high levels of performance not only by applying proper platform-specific code optimizations (see Sect. 4), but also by choosing hardware-friendly operations in this most crucial algorithmic core operation.

Obviously, the straight forward solution for computing the numeric maximum of two pixel values involves some branch-instruction. Probably all recent high-performance CPU-hardware offer intricate and even online performance optimization techniques, such as instruction reordering, branch prediction, and speculative execution. Along with hierarchical multi-level caching memory CPUs are mostly capable of executing branching operators without suffering from significant performance penalties. The situation is quite different on many modern accelerator-based hardware platforms, which usually offer higher compute core counts at the cost of reduced core complexity. Our rationale here is to avoid branching operations to a high degree, as these operations tend to stall the stream of instructions on GPGPU-hardware, which diminishes overall throughput. Also, on hardware that supports true vector-processing[4] such as the high-performance computing platform NEC ACE SX, a steady stream of branch-less instructions promises to be beneficial towards our goal of high computational throughput.

[4]While all modern CPUs do actually support high-throughput instructions that operate on short vectors of data elements (e.g. SSE, AVX, Altivec, etc.), we want to make the distinction against pipelined vector processing, which is capable of processing vectors of arbitrary length while also employing a richer set of instructions compared to standard x86-based processors.

```
#define MAX(a,b)      (((a)>(b))?(a):(b))
```

**Fig. 1** Branch-based maximum operator

```
#define MAX(a,b)      (a-((a-b)&(a-b)>>31))
```

**Fig. 2** Closed-form maximum operator

### 3.2.2 Branch-Based Maximum Operator

The definition depicted in Fig. 1 constitutes a classic macro of the C language, which translates into efficient code on modern CPU-hardware, such as Intel x86 or IBM Power architectures.

### 3.2.3 Closed-Form Maximum Operator

Figure 2 constitutes the closed-form macro[5] for computing the maximum of two signed integer values (int32 data type). It involves no branching operation, but basic arithmetic and bit-wise operations only. Due to the absence of branch operators, this function incurs no warp-divergence on CUDA-enabled devices and promises benefits on any in-order execution compute platform.

Finally, a host-based driver routine (Algorithm 4) is used to orchestrate the series of compute and termination criterion (Algorithm 2) functions that resemble the Cellular Automaton.

Considering the GPU implementation, updates of the CA and corresponding termination checks will exclusively be accomplished in GPU RAM.

**Algorithm 4** Driver

1: **procedure** DRIVER(input: binary lattice, output: CA1)
2:     *CA1* ← *initial coloring (binary lattice)*
3:     *CA2* ← *CA1*
4:     **repeat**
5:         *CA_update*(*CA*1, *CA*2)
6:         *CA_update*(*CA*2, *CA*1)
7:     **until** termination

[5] As proposed by Holger Berger of NEC Germany.

# 4 Implementation

Our baseline implementation consists of OpenMP-enhanced x86-code (C language), from which several code-branches for GPU (nVidia CUDA), multi-GPU, Intel Xeon Phi, and NEC ACE code have been derived. These approaches support dense data sets, which are stored as a standard array in C. Depending on density and distribution of non-background pixels in a given data set, we find that an alternate, sparse representation of Cellular Automaton data offers performance benefits, albeit at the cost of some increase in memory usage.

For performance reasons we decided to employ a granularity of two updates per termination check. The main advantage of this design decision is the potential for omitting any swapping operations on input and output state arrays, such as described in [3]. By switching input- and output parameters between two consecutive calls to the update function, state arrays CA1 and CA2 serve both as input and output array. The frequency of calls to the termination criterion function is also reduced as a consequence, which is preferable for reasons of performance but in general will also lead to one potentially superfluous update operation in the last phase of the algorithm as it reaches the equilibrium state of the CA.

## *4.1 Dense Data Representation*

## *4.2 OpenMP-Code*

OpenMP is the industry-standard for task-level parallelization on multi-processor systems. It allows for convenient development cycles, which usually start from a sequential code base. By incrementally adding parallel constructs to the code, the execution speed will be enhanced and the software will be enabled to utilize all available system resources. Fortunately, NEC is offering their own high-performance implementation of the OpenMP runtime and compiler environment so that porting efforts starting from the x86-based code base prove to be a rather straight-forward process. Consequently, the resulting source codes both for x86- and NEC ACE-SX systems do look very similar and we only want to give a glimpse on the differences.[6] Porting the core update function to the Xeon Phi (KNC) processor follows Intel's standard programming model called 'function offloading'. In this model the Intel C-compiler is guided by means of a few directives to generate parallel OpenMP-based code for the Xeon Phi-accelerator as well as the necessary staging of function data (e.g. state arrays CA1, CA2). Figure 3 displays the x86-based update routine.

[6]The NEC implementation of OpenMP offers pragma based hints to the compiler which signify independence of nested loops, such as loops 'row' and 'col' in Fig. 3. By adding '#pragma cdir nodep' to the inner loops, the compiler is set to optimize in more aggressive way.

```
void update_CPU (int N, int *input, int *output)
{
int row, col, slice;
int cell;

#pragma omp parallel for private (row, col, cell)
for (slice = 1; slice < SDIM -1; slice++)
  {
  for (row = 1; row < SDIM -1; row++)
    {
    for (col = 1; col < SDIM -1; col++)
      {
      if (input[slice * N * N + (row * N) + col] != 0)
        {
        cell = input[slice * N * N + ((row) * N) + col];
        cell = Max(cell, input[slice * N * N + ((row-1) * N) + col]);
        cell = Max(cell, input[slice * N * N + (row * N) + col-1]);
        cell = Max(cell, input[slice * N * N + (row * N) + col+1]);
        cell = Max(cell, input[slice * N * N + ((row+1) * N) + col]);
        cell = Max(cell, input[(slice-1) * N * N + (row * N) + col]);
        cell = Max(cell, input[(slice+1) * N * N + (row * N) + col]);
        output[slice * N * N + (row * N) + col] = cell;
        }
      }
}}}
```

**Fig. 3** OpenMP code: parallel update of CA cells

## *4.3 CUDA Implementation*

For achieving high computational performance in the Cellular Automaton kernel we find two very relevant design decisions. First, data decomposition has to fit the memory subsystem of the GPU hardware. This essentially boils down to proper memory coalescing, which is a standard technique of forcing adjacent CUDA cores access adjacent memory locations in parallel. We aim to achieve high memory throughput on the GPU by assigning an appropriate number of CUDA threads to the computation of the inner-most loop ('columns') (4), which as a result displays the necessary memory access patterns. In other words, the inner-most loop is squashed altogether and being replaced by an appropriate number of CUDA threads that operate in parallel. This limits the maximum dimension of the CA to the maximum number of CUDA threads per CUDA block. For current generation NVIDIA-hardware this amounts to 1024 threads, hence a maximum dimension of $1024^3$ cell elements[7] is being supported by our current GPU-implementation.

[7]While this may seem to be a limiting factor in the application of the kernel for large CAs, it should be stated that the corresponding amount of necessary GPU-memory quickly fills the available on-chip resources of the accelerator, which might be the main limitation towards employing larger datasets.

The second design decision of the CUDA kernel involves the method of parallelizing the two outer loops ('rows', 'planes') of the kernel. Since no memory-coalescing issues have to be taken into account at this level, we enjoy freedom to employ loops, a CUDA grid-based decomposition, or some mixture of both design models. Relying on for-loops only puts high computational pressure on each CUDA thread and—probably even more important—hampers the inherent capability of the GPU in hiding latencies of memory accesses by employing large numbers of active CUDA blocks and threads. Also, due to the very high core counts (e.g. 3584 cores on recent PASCAL-cards) of modern high performance hardware some loop-only based approach would severely limit the degree of parallelism.[8]

Historically, CUDA-enabled devices offered little or no L2-cache memory, but some fast 'shared memory' or scratch-pad memory on-chip. By resorting to software-controlled caching mechanisms this lack of hardware-controlled L2-cache could in general be alleviated by clever kernel design. Recent generations of CUDA-hardware do offer improvements both in terms of size and levels of control of cache memory. Nevertheless, there is no easy way to decide whether to just rely on hardware-controlled L2-caching mechanisms or to exert explicit control over memory access by resorting to 'old-style' programming techniques.

In order to achieve maximum performance we apply explicit cache control by way of memory-coalescing during column-reads (function 'read_column' in Fig. 4) in conjunction with implicit, hardware-based cache control. As can also be seen in Fig. 4, the number of column-reads can be diminished for all row iterations except for the first one. This is accomplished by explicitly copying column data that is already present in the shared memory segment 'local' following the direction of the loop. Hence, for each new iteration in which the stencil-column is being moved towards the last row of the current slice ($z$-plane of the CA), three instead of five column-reads are sufficient, which marks a 40% reduction of memory transfers.

### 4.3.1 Termination Criterion

Checking the termination criterion in parallel is based on finding any discrepancy between state arrays CA1 and CA2. Depending on the dimensions of the computed CA, this involves transfer of data on the order of multiple GiB, which makes repeated checks prone to becoming a major bottleneck of the algorithm. We therefore aim for an early termination of the check routine itself, which requires fast synchronization of collaborating GPU-threads. As outlined in code 'termination', read access to both state arrays CA and CA2 is being coalesced by enabling threads to work in lockstep on properly aligned data. Discrepancies within two given data columns lead to immediate termination of those threads, that spotted some discrepancy. Equally important, discrepancies will raise some global flag, which

---

[8]Employing 3584 cores to a CA-kernel of dimension $1024^3$ would yield hardware utilization rates below of 29%.

```
__global__ void update_GPU ( const int * __restrict__ input,
                             int * __restrict__ output)
{
__shared__ int local[6][N];   // N = dimension of CA
int slice = blockIdx.x;
int row, col = threadIdx.x;
int max;

if ((slice >0) {

for (row = 1; row < (N-1); row++)
   {
    __syncthreads();
   if (row==1)  // read full stencil
     {
     readcolumn (&local[0][0], &input[slice*N*N+(row-1)*N]);
     readcolumn (&local[1][0], &input[slice*N*N+row*N]);
     readcolumn (&local[2][0], &input[slice*N*N+(row+1)*N]);

     readcolumn (&local[3][0], &input[(slice-1)*N*N+row*N]);
     readcolumn (&local[4][0], &input[(slice+1)*N*N+row*N]);
     local[5][threadIdx.x]=local[1][threadIdx.x];//output line
     }
    else     // read partial stencil with reuse of recent data
     {
     readcolumn (&local[3][0], &input[(slice-1)*N*N+row*N]);
     readcolumn (&local[4][0], &input[(slice+1)*N*N+row*N]);
     local[0][threadIdx.x] = local[1][threadIdx.x]; // reuse
     local[1][threadIdx.x] = local[2][threadIdx.x]; // reuse
     readcolumn (&local[2][0], &input[slice*N*N +(row+1)*N]);

     local[5][threadIdx.x]=local[1][threadIdx.x];//output line
     }

  __syncthreads(); // wait for data to have arrived
   if ((threadIdx.x > 0)&&(threadIdx.x < N-1)) {
     if (local[1][threadIdx.x] != 0)
     {
     max=local[0][threadIdx.x]>local[1][threadIdx.x] ? _
           local[0][threadIdx.x] : local[1][threadIdx.x];
     max=local[1][threadIdx.x-1]>max?local[1][threadIdx.x-1]:max;
     max=local[1][threadIdx.x+1]>max?local[1][threadIdx.x+1]:max;
     max=local[2][threadIdx.x]>max ? local[2][threadIdx.x] : max;
     max=local[3][threadIdx.x]>max ? local[3][threadIdx.x] : max;
     max=local[4][threadIdx.x]>max ? local[4][threadIdx.x] : max;

     local[5][threadIdx.x] = local[4][threadIdx.x]>max ? _
                             local[4][threadIdx.x] : max;
     }
   }
   // store resulting column in global output array
   writecolumn (&local[5][0], &output[slice * N * N + row * N]);
  }
 } // slice
}
```

**Fig. 4** CUDA code: parallel evaluation of termination criterion

will prevent any not-yet active CUDA block from entering the checking routines. It has to be noted that thread termination within some active CUDA block will only affect remaining threads of that block. This might be regarded to be sub-optimal, but experiments with an increased level of synchronization at Warp-level exhibited inferior overall performance.

Please also note the absence of any explicit synchronization construct in the above implementation of non-blocking synchronization.

Interestingly enough, efficient parallelization of the termination criterion proves to be much harder in OpenMP than in CUDA, due to the absence of premature termination of parallel for-loops in OpenMP. As a consequence, OpenMP code will be forced to sift through both state arrays CA1 and CA2 in total, even when differences should have been spotted during the first few comparisons. Albeit explicit and more complex task-based implementations within OpenMP would have been possible, we instead opt for a sequential version. Due to the simplicity of the core operator (check for equality of two cell states) the resulting code will operate close to the saturation level of the memory system, which can be taken as an argument against parallelization of this code section in the first place (Fig. 5).

```
__global__ void termination (const int N,
                   const int * inputA, const int * inputB,
                   unsigned char * __restrict__ activity)

{
__shared__ int local[2][SDIM];

int slice = blockIdx.x;
int row;
int col = threadIdx.x;
int max;
int p;
unsigned char flag=0;

if ((slice>0)&&(slice<SDIM-1)) {
  if (activity[0]==0) {
   for (row = 1; row < (N-1); row++)
     {
     readcolumn (&local[0][0], &inputA[slice * N * N + row * N]);
     readcolumn (&local[1][0], &inputB[slice * N * N + row * N]);

     if (local[0][threadIdx.x] != local[1][threadIdx.x])
        {
        row=N;        // terminate thread
        activity[0]=1;  // raise global termination flag
        }
     }
   } // if activity==0
  } // if slice > 0
}
```

**Fig. 5** CUDA code: parallel evaluation of termination criterion

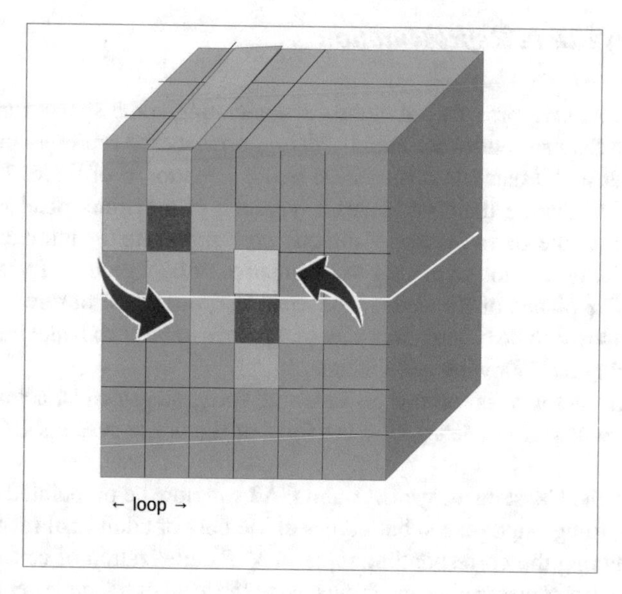


**Fig. 6** In dual-GPU environments, state arrays are partitioned into two even portions with each portion being stored locally on one of the participating GPUs. Both GPUs may access state arrays of the corresponding partner GPU by means of unified virtual addressing (UVA) mode

## *4.4 Multi-GPU Computation*

Figure 6 depicts the basic layout of CA data in dual-GPU setups. Each GPU is enabled to access ghost-cells (cells that are read but never being written to) that physically reside in the partner GPU's local memory by means of unified virtual addressing (UVA) mode. However, the CUDA kernel is repeatedly forced to decide whether a certain column of data is available locally or whether it has to be fetched via the UVA mechanism. This decision adds to an increase in code complexity[9] and potentially also harms the overall throughput of the kernel. By introducing two separate kernels that are specialized for operation in the areas of ghost cells that emerge at the lower and upper borders of their data partition, we aim to alleviate this performance bottleneck. The performance numbers reported on in the following section are based on this improved multi-kernel model.

[9]While code complexity is not regarded an issue on standard CPU-based systems, it certainly can lead to an inflation of the size of the binary executable, which in extreme cases can result in non-executable kernels.

## *4.5 Sparse Data Representation*

For easier usage we provide a method for generating some sparse representation of any given data set out of its initially dense, array-based representation. For the discussed dense 3D datasets this method builds a vector V of tuples $(x, y, z)$ with each tuple designating the coordinates $x$, $y$, and $z$ of a distinct pixel in the dense data set. Hence, the size of vector V directly corresponds to the number of relevant pixels, that is, pixels not belonging to boundaries or background. By cycling over the $z$, $y$, and $x$-planes of the dense dataset in that order, we ensure the tuples of resulting vector V to be ordered in a way that proves to be cache- and memory-page friendly during the following cell update.

Note that vector V is merely an index of foreground-pixels, actual CA state information will still be managed in the form of the dense systems CA1 and CA2 (Sect. 4.1).

Updating the CA state arrays CA1 and CA2 can now be pinpointed to the exact locations of foreground pixels, but comes at the cost of additional memory access for dereferencing the corresponding tuple in V. Parallelization of cell updates is a straight-forward process being accomplished at the level of the tuple vector V, which is a densely packed dataset that lends itself well to OpenMP-, CUDA-, and vector processing approaches. Nevertheless, at this point the general choice of dense versus sparse data representation has to be grounded on heuristics.

#### 4.5.1 OpenMP-Code

The structure of nested loops in the dense case (as shown in Fig. 3) is replaced by a single loop (Fig. 7) that operates on the vector V of tuples.

## 5 Simulation Results

The simulation code for the CA has been run under benchmarking conditions on a set of systems with the intention of giving 'the bigger picture' on what performance levels are to be expected on recent high-performance compute hardware. With the notable exception of the NEC ACE-SX system, the employed hardware belongs to the class of so-called accelerators that—probably also due to its potential performance and affordability—seems to be attaining more attention both in science and engineering for quite some time now. **Table 3 gives an overview of obtained speedups of the investigated compute architectures over the baseline system Intel Xeon 1620 (quad core)**. Again, we want to stress the fact that these

```
void update_CPU_sparse (int N, int *input, aPixel *V,
                        unsigned int pixels, int *output)
{
int row, col, slice;
int cell;
unsigned int nr;

#pragma omp parallel for private (slice, row, col, cell)
for (nr =0; nr < pixels; nr++)
   {
   slice = V[nr].z;
     row = V[nr].y;
     col = V[nr].x;

   cell = input[slice * N * N + ((row) * N) + col];
   cell = Max(cell, input[slice * N * N + ((row-1) * N) + col]);
   cell = Max(cell, input[slice * N * N + (row * N) + col -1]);
   cell = Max(cell, input[slice * N * N + (row * N) + col+1]);
   cell = Max(cell, input[slice * N * N + ((row+1) * N) + col]);
   cell = Max(cell, input[(slice -1) * N * N + (row * N) + col]);
   cell = Max(cell, input[(slice+1) * N * N + (row * N) + col]);
   output[slice * N * N + (row * N) + col] = cell;
   }
}
```

**Fig. 7** OpenMP code: evaluation of sparse state array

**Table 3** Speedup overview (based on results presented in Table 1)

| CA dim | GTX680 | NEC ACE-SX | Tesla P100 | Xeon E1620 | Xeon Phi 5120 |
|---|---|---|---|---|---|
| 128 | 45× | 19.9× | 43.7× | 1× | n/a |
| 256 | 37× | 18.5× | 99.7× | 1× | n/a |
| 512 | 25.7× | 12.5× | 68.9× | 1× | n/a |
| 704 | 15.5× | 9.7× | 55.2× | 1× | 18.1× |

systems stem from different cycles in hardware development, so the interpretation of reported numbers should take that into account, too. The Pascal-based Tesla P100 offers extreme levels of performance with runners-up found in the GTX680 GPU and the Xeon Phi. In light of the high thread-counts of these three platforms which range from several hundreds up to more than 3.500 threads the performance of the quad core vector processor ACE-SX is quite astonishing (Fig. 8).[10]

Figure 9 depicts the relation of execution times for dense and sparse codes on two hardware platforms, one GPU and one node of the NEC ACE-SX vector

[10]Even more so when we want to put this into relation with the age of the hardware concept of this generation of the NEC processor, that apparently goes back at least 5 years from the time of this report.

```
__global__ void sparse_update (unsigned int number,
                               const aPixel* __restrict__ pixel_list,
                               const int * __restrict__ input,
                               int * __restrict__ output)

{
int max, iter, x,y,z;
unsigned int pos;

pos = (THREADS * Block_LOOPS)*blockIdx.x+threadIdx.x;

for (iter = 0; iter < Block_LOOPS; iter++)
 {
 if (pos < number)
  {
  z=(int)pixel_list[pos].z;
  y=(int)pixel_list[pos].y;
  x=(int)pixel_list[pos].x;

  max=input[z*N*N+(y-1)*N+x] > input[z*N*N+y*N+x] ? _
      input[z*N*N+(y-1)*N+x] : input[z*N*N+y*N+x];
  max=input[z*N*N+y*N+x-1]>max ? input[z*N*N+y*N+x-1]:max;
  max=input[z*N*N+y*N+x+1]>max ? input[z*N*N+y*N+x+1]:max;
  max=input[z*N*N+(y+1)*N+x]>max ? input[z*N*N+(y+1)*N+x]:max;
  max=input[(z-1)*N*N+y*N+x]>max ? input[(z-1)*N*N+y*N+x]:max;

  output[z*N*N+y*N+x] = input[(z+1)*N*N+y*N+x]>max ? _
                        input[(z+1)*N*N+y*N+x]:max;

  pos += (THREADS);
  } // pos < number
 }
}
```

**Fig. 8** CUDA code: evaluation of sparse state array

processor system, respectively.[11] Both platforms deliver very stable execution times for updates of the dense CA. As expected, execution times in the sparse formulation of the CA update compare favorably to their dense counterparts for highly sparse systems (e.g. upwards of 95% of background pixels). As can be seen, the GPU system performs in robust way after reaching saturation at sparsity levels in the range of 10–15%, whereas the vector processor ACE-SX seems to struggle with the increased level of scattered memory access. It has to be noted that ACE-SX reaches the cut-off point at which the sparse code no longer prevails over the dense code at a later stage than the GPU platform. Hence, sparse formulation of updates is beneficial for a wider range of densities in a given CA on ACE than it is for the GPU.

[11]The results that we report here have to be taken with some caution, as the ACE-SX and GTX 1080Ti belong to rather different eras of their respective development time.

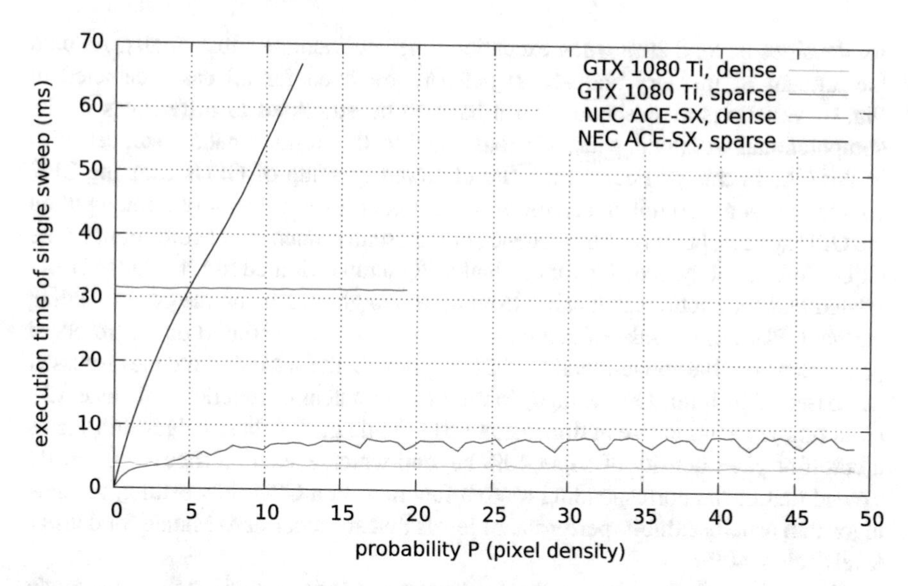


**Fig. 9** Comparison of NVIDIA GTX1080 Ti GPU with NEC ACE-SX for dense and sparse systems of dimension 512

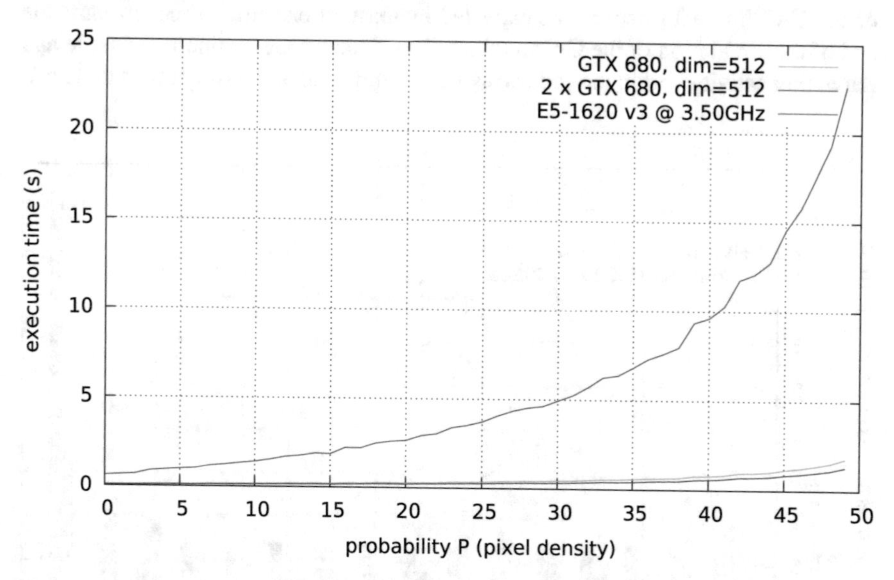


**Fig. 10** Relation of runtime and pixel probability (density of CA) for single-, dual-GPU and quad-core CPU systems

As shown in Fig. 10, both GPU setups (single GTX 680, dual GTX 680) exhibit performance levels that are robust against increased levels of pixel densities of the simulated CA. On the contrary, the Xeon E5-1620-based system is able to maintain relatively low turn-around times for low pixel densities, but falls short for densities beyond 20% when execution times do increase substantially. As both the amount of memory and access patterns are fixed for all cases depicted in Fig. 10 variations of execution times have to be attributed to differences of the computational workload, which obviously is directly proportional to pixel densities in the CA. In this particular case, the observed speedup of GPUs over the CPU counterpart is not so much a result of some higher level in memory throughput on the GPU system, but it is a consequence of the usually much lower core count on the CPU. Unfortunately, speedup for the dual-GPU setup is limited to 1.4× due to UVA-related PCI transactions that result from access to ghost cells on the corresponding partner GPU. For the sake of completeness, we also want to report on the observed performance on the above-mentioned CPU system both for Matlab and the discussed CA-based algorithm. On average, Matlab's bwconncomp function will take 12 s (wall time) in a 3-d lattice of dimension=512, while the CA-based algorithm shows a cutoff at pixel density of some 40% beyond which execution times will mostly exceed that of the corresponding Matlab function. In a CPU-only setting, the new algorithm usually exhibits performance levels that are superior to Matlab for density levels below 40%.

The optimized checking of the termination criterion yields stable and much reduced turn-around times or single iterations up to some 70% of the total runtime of the CA. Figure 11 depicts the expected increase of execution times towards the end of the simulation of the CA, in which the CA states are beginning to settle and remaining activity within the state array CA1 requires an increasing effort to detect.

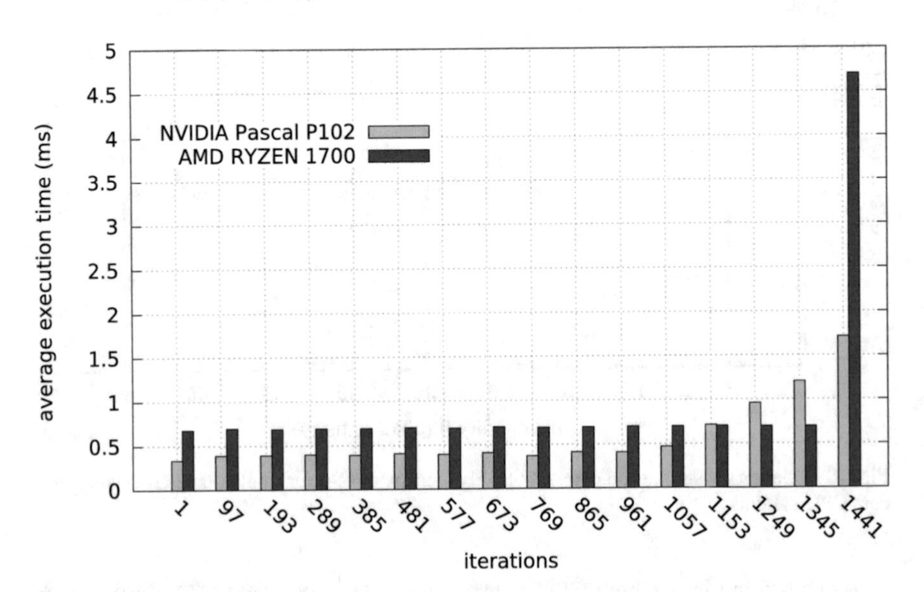


**Fig. 11** Execution timings of termination checks (based on code given in Fig. 5)

## 5.1 Maximum Operator

Our initial motivation for conducting experiments with an alternative, closed-form maximum operator has been driven by the typical hardware characteristics of modern GPUs, that on one hand do offer numerous numbers of powerful CUDA cores, but on the other hand suffer from branching-incurred penalties (e.g. warp-diversion). Unfortunately, the increased computational complexity of the closed-form operator cannot be overcome by the sheer power of the GPU system.

As displayed in Fig. 12 an approach employing the branch-based maximum-operator (Fig. 1) in conjunction with a loop-based computation of the row elements of the CA state array performs best on the NVIDIA GTX 780 Ti hardware. Depending on the dimensions of the CA the next best option for dimension=256 is stated by a CUDA grid-based parallelization of the row elements and for dimension=512 the second best option is given by a combination of loops and closed-form maximum operator. This difference can be attributed to the increased overhead that is introduced by the large numbers of CUDA blocks in dimension=512 which amount to $512^2 = 262{,}144$. Apparently, instantiation of such large numbers of CUDA-blocks happens to be a rather expensive task in itself. Based on this observation a second important finding follows, which can be stated as the occasional necessity

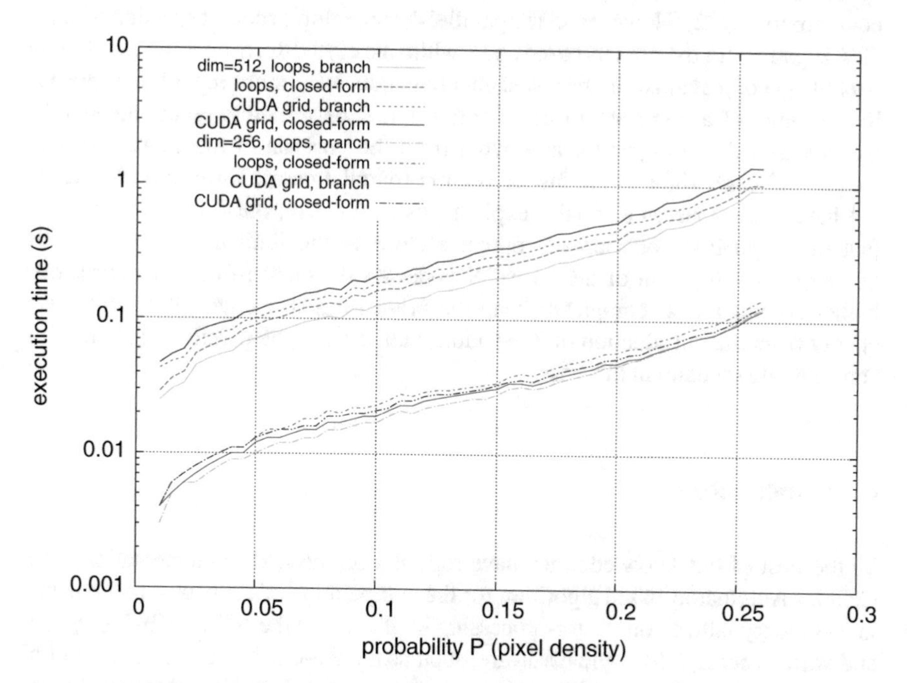


**Fig. 12** NVIDIA GTX 780Ti: various update kernels based on loops vs. CUDA grid and branching vs. closed-form maximum operators

**Table 4** Performance and scalability on NEC ACE-SX

| Max operator | CPU core(s) | Execution (ms) | Speedup | Global speedup |
|---|---|---|---|---|
| branch-based | 1 | 107.9 | 1.0 | 1.0 |
| closed-form | 1 | 233.11 | 0.46× | 0.46× |
| branch-based | 4 (OpenMP) | 81.74 | 1.0 | 1.32× |
| closed-form | 4 (OpenMP) | 79.03 | 1.034× | 1.37× |

for conventional loop-structures on GPUs, even when CUDA allows for a clean and lean formulation of an algorithm entirely devoid of loops.

On NEC ACE-SX an ambivalent situation prevails when the closed-form maximum operator is applied. This system consists of a number of distributed cluster-nodes, that communicate via a high-speed network and message-passing style communication primitives. Each cluster-node comprises four distinct vector processing cores, that allow for very convenient and tightly coupled OpenMP-based parallelization as well as vectorization within the cores. Our experimental setup concentrates on a single such node, hence we may apply vectorization and shared-memory parallelization, respectively. In the single core variant the update of the state array suffers from severe performance penalties (see Table 4, rows 1 and 2) when the standard branch-based maximum operator (Fig. 1) is replaced by its closed-form counterpart (Fig. 2). However, in the parallel 4-core setup a modest speedup of some 32% is gained for the branch-based code, while the closed-form operator yields 37% overall speedup against the fastest sequential code. We do not regard this speedup in the range of a mere 3% to be essential, but it nevertheless seems noteworthy that the closed-form operator is moving from the position of the slowest version (sequential case) right to the fastest version (parallel case). Unfortunately, we do not have the resources to finally explain this result here, but it can be suspected that the sequential code already operates close to the limit that is imposed by the memory subsystem of the ACE-SX node. As the closed-form operator sports higher computational complexity than the branch-based operator, it benefits very clearly from the introduction of three additional cores in the parallel run, hence the considerable speedup in this case.

## 6 Conclusions

To the best of our knowledge we have reported on the first implementations of a Cellular Automaton-based algorithm for the computation of connected components in 3-d binary lattices on vector-processing hardware, as the NEC ACE-SX system and various accelerator-style hardware, such as nVidia-CUDA- and Intel Xeon Phi equipped systems. The algorithm proves to be very suitable to all of these platforms, and high levels of performance have been gained due to the massive amounts of parallelism that is inherent to this class of algorithms. Even the baseline code that we benchmarked on some modest mainstream Intel Xeon CPU offers certain performance benefits in comparison to the corresponding function that is provided with the popular Matlab environment. The reported performance numbers clearly indicate potential benefits from experimenting with different algorithmic solutions, even within the same general platforms (e.g. GPU).

**Acknowledgements** We would like to sincerely thank Professor Michael M. Resch and the whole team of HLRS for their valuable support, continued guidance and discussions, and provision of systems.

## References

1. Datta, K., et al.: Stencil computation optimization and auto-tuning on state-of-the-art multicore architectures. In: Proceedings of the 2008 ACM/IEEE Conference on Supercomputing (2008)
2. Hoekstra, A.G., Kroc, J., Sloot, P.M.A.: Simulating Complex Systems by Cellular Automata. Springer, Berlin (2010)
3. Holewinski, J., Pouchet, L.-N., Sadayappan, P.: High-performance Code Generation for Stencil Computations on GPU Architectures. ACM, New York (2012). doi:10.1145/2304576.2304619
4. Stamatovic, B., Trobec, R.: Cellular automata labeling of connected components in n-dimensional binary lattices. J. Supercomput. **72**(11), 4221–4232 (2016). doi:10.1007/s11227-016-1761-4
5. Tang, Y., Chowdhury, R.A., Kuszmaul, B.C., Luk, C.-K., Leiserson, C.E., The Pochoir Stencil Compiler. ACM, New York (2011). doi:10.1145/1989493.1989508
6. Trobec, R., Stamatovic, B.: Analysis and classification of flow-carrying backbones in two-dimensional lattices. Adv. Eng. Softw. **103**, 38–45 (2015)

# Part IV
# Computational Fluid Dynamics

# Turbulence in a Fluid Stratified by a High Prandtl-Number Scalar

**Shinya Okino and Hideshi Hanazaki**

**Abstract** Turbulence in the fluid stratified by a high Prandtl-number ($Pr$) scalar such as heat ($Pr = 7$) or salinity ($Pr = 700$) has been simulated by direct numerical simulations, using $4096^3$ grid points. Computations have been performed using the 1024 nodes of NEC SX-ACE, which have enabled us to resolve the smallest scale of salinity fluctuations $\sqrt{700}(\sim 26)$ times smaller than the smallest eddy size. In our simulations, buoyancy initially affects only the large scale motions, and the $k^{-1}$ spectrum predicted by Batchelor (J. Fluid Mech. 5:113–133, 1959) for a passive scalar could be observed in the spectrum of potential energy, i.e. the salinity fluctuations. However, as time proceeds, the buoyancy affects the smaller-scale motions, and the salinity fluctuations begin to show a unique spatially localised structure. At the same time, there appears a flat spectrum ($\propto k^0$) instead of the $k^{-1}$ spectrum. The localised structure and the flat spectrum could be observed only for the salinity ($Pr = 700$) and not for heat ($Pr = 7$).

## 1 Introduction

The atmosphere and the ocean, when temporally averaged, are the stably stratified fluids with larger density at lower altitude. The oceanic flow is determined by the distribution of temperature and salinity, but the salinity has a very small diffusion coefficient so that the Prandtl number $Pr$, which is the ratio of the kinematic viscosity of fluid to the diffusion coefficient, is very large ($Pr = 700$). This means that the smallest scale of salinity fluctuations is much smaller than the smallest eddy size in the flow. Batchelor [1] studied the variance spectrum of the passive scalar which is simply convected by the flow without exerting the buoyancy effect on the fluid. He showed theoretically that a high Prandtl number scalar ($Pr \gg 1$)

S. Okino • H. Hanazaki (✉)
Department of Mechanical Engineering and Science, Graduate School of Engineering, Kyoto University, Kyoto daigaku-katsura 4, Nishikyo-ku, Kyoto 615-8540, Japan
e-mail: okino@me.kyoto-u.ac.jp; hanazaki.hideshi.5w@kyoto-u.ac.jp


M.M. Resch et al. (eds.), *Sustained Simulation Performance 2017*,
DOI 10.1007/978-3-319-66896-3_7

dissipates at the wavenumber $k_{B*} = Pr^{1/2}k_{K*}$ ($k_*$: the Kolmogorov wavenumber), which is now called the Batchelor wavenumber, and the spectrum is proportional to $k^{-1}$ in the viscous-convective subrange ($k_{K*} < k_* < k_{B*}$).

Batchelor's prediction has been confirmed by experiments [2], but the corresponding numerical simulation of turbulence with a high Prandtl-number scalar is difficult since the Batchelor wavenumber increases in proportion to $Pr^{1/2}$. For example, the smallest scale of salinity fluctuation ($Pr = 700$) is $\sqrt{700}(\sim 26)$ times smaller than the Kolmogorov scale, which is the smallest scale of the fluid motion. This indicates that the computation of salinity distribution requires $26^3(\sim 20{,}000)$ times as many grid points as the turbulent flow itself. While this difficulty remains, high Prandtl-number passive scalars ($Pr > 1$) in homogeneous fluids have been simulated numerically to validate Batchelor's prediction (e.g. [3]), and the turbulent mixing of a very high Prandtl-number passive scalar ($Pr = O(10^3)$) has been investigated [4].

As for the stratified turbulence with a buoyant scalar (e.g. [5, 6]), there have been many numerical studies, but almost all of them assume $Pr \sim 1$ to avoid the difficulty of resolution. Therefore, behaviour of the high-$Pr$ buoyant scalar and its effect on the fluid motion have been largely unknown. In this study, we perform the direct numerical simulations of decaying turbulence in the fluid stratified by a high Prandtl-number scalar such as heat ($Pr = 7$) or salinity ($Pr = 700 \gg 1$). The results for a very-high Prandtl number scalar ($Pr = 700$) would be particularly useful to understand the mechanisms underlying the results of laboratory experiments, or to consider the final period of decay of small-scale turbulence in the real ocean.

## 2 Direct Numerical Simulations

We consider the motion of an incompressible fluid in a periodic cube $4\pi L_{0*}$ ($L_{0*}$: initial integral scale) on each side. The fluid is uniformly stratified in the vertical direction by scalar $\tilde{T}_*$, and the Brunt-Väisälä frequency $N_*$ is defined by $N_* = \sqrt{-\alpha_* g_* d\overline{T_*}/dz_*}$, where $\alpha_*$ is the contraction coefficient of the scalar, $g_*$ is the gravitational acceleration and $d\overline{T_*}(z_*)/dz_*$ is the constant vertical gradient of the undisturbed scalar distribution. We investigate the decaying turbulence in a stratified fluid. The initial velocity fluctuation is isotropic with rms velocity $U_{0*}$ and integral length $L_{0*}$, while the initial scalar fluctuation is assumed to be absent.

The temporal variation of the velocity $u_{i*}$ ($i = 3$ denotes the vertical component) and the scalar deviation from its background value $T_* = \tilde{T}_* - \overline{T_*}(z_*)$ are governed by the Navier-Stokes equations under Boussinesq approximation, the scalar-transport equation and the incompressibility condition:

$$\frac{\partial u_i}{\partial t} + u_j \frac{\partial u_i}{\partial x_j} = -\frac{\partial p}{\partial x_i} + \frac{1}{Re_0}\frac{\partial^2 u_i}{\partial x_j^2} - \frac{1}{Fr_0^2} T \delta_{i3}, \tag{1}$$

$$\frac{\partial T}{\partial t} + u_j \frac{\partial T}{\partial x_j} = \frac{1}{Re_0 Pr}\frac{\partial^2 T}{\partial x_j^2} + u_3, \tag{2}$$

$$\frac{\partial u_i}{\partial x_i} = 0, \tag{3}$$

where the physical quantities without an asterisk are non-dimensionalised by the length scale $L_{0*}$, the velocity scale $U_{0*}$ and the scalar scale $T_{0*} = -L_{0*} d\overline{T_*}/dz_*$. Non-dimensional parameters of the system are the initial Reynolds number $Re_0 = U_{0*}L_{0*}/\nu_*$, the initial Froude number $Fr_0 = U_{0*}/N_* L_{0*}$ and the Prandtl number $Pr = \nu_*/\kappa_*$, where $\nu_*$ is the kinematic viscosity coefficient and $\kappa_*$ is the diffusion coefficient of the scalar. In this study, the initial Reynolds number and the initial Froude number are fixed at $Re_0 = 50$ and $Fr_0 = 1$, while two Prandtl numbers $Pr = 7$ and 700 are considered, corresponding to the heat and salinity diffusion in water.

The governing equations are solved by the Fourier spectral method. The nonlinear terms are evaluated pseudospectrally and the aliasing errors are removed by the 3/2-rule. As the time-stepping algorithm, the 4th-order Runge-Kutta method is used. The initial fluctuation is developed without stratification until the enstrophy reaches its maximum, and the stratification is switched on at that moment which is defined as $t = 0$, and the computation is continued until $t = 30$.

Since the initial Reynolds number and the initial Froude number are fixed ($Re_0 = 50$, $Fr_0 = 1$), the initial Kolmogorov wavenumber is also a constant ($\sim$17), and the Batchelor wavenumber for $Pr = 700$ is $17 \times \sqrt{700} \sim 450$. Such a small scale of scalar fluctuation can be resolved if we use $4096^3$ grid points, since the maximum wavenumber becomes 682.5 when the minimum wavenumber is 0.5 (note that the flow is spatially periodic with the period of $4\pi$ in the non-dimensional form). Since the high-wavenumber components decay with time, we have reduced the number of grid points to $2048^3$ when the Batchelor wavenumber becomes smaller than approximately 210 ($t = 6$), in order to save the computational resources.

The numerical simulation is executed using the 1024 nodes of NEC SX-ACE in Tohoku University and Earth Simulator Center of JAMSTEC. Since more than 90% of the computational time is spent for the three-dimensional real FFT to evaluate the nonlinear terms, we have parallelised it based on the 1D-decomposition, using the Message Passing Interface (MPI). The parallelised code showed a good scalability up to 1024 nodes (Fig. 1).

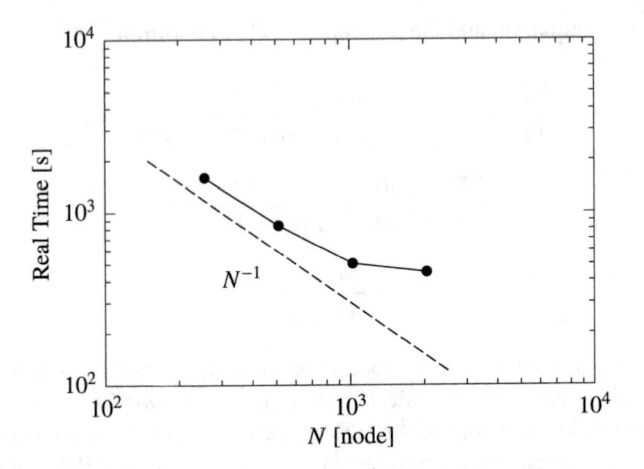


**Fig. 1** The relation between the number of nodes ($N$) used for the computation and the real time necessary for advancing 10 time steps when the number of grid point is $4096^3$. The line of $N^{-1}$ (ideal scalability) is drawn for reference

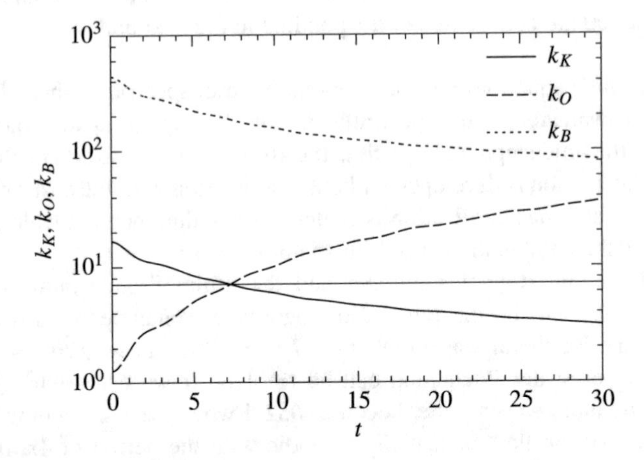


**Fig. 2** Temporal variation of the Kolmogorov wavenumber $k_K$, the Ozmidov wavenumber $k_O$ and the Batchelor wavenumber $k_B$ for $Pr = 700$. When the Kolmogorov wavenumber agrees with the Ozmidov wavenumber, $k_K = k_O = (Re_0/Fr_0)^{1/2} \sim 7.1$

## 3 Results

We first show in Fig. 2 the temporal variation of the characteristic wavenumbers for $Pr = 700$. The (dimensional) Kolmogorov wavenumber defined by $k_{K*} = (\epsilon_*/\nu_*^3)^{1/4}$ ($\epsilon_*$: kinetic energy dissipation rate) corresponds to the smallest size of vortices, and it decreases with time because small scale fluctuations decay as time proceeds. The Ozmidov wavenumber is defined by $k_{O*} = (N_*^3/\epsilon_*)^{1/2}$, estimating the wavenumber at which the inertial force and the buoyancy force balance. It is initially one-order of magnitude smaller than the Kolmogorov wavenumber and increases with time, meaning that the buoyancy effect is initially limited to the large scale motion, and smaller scales are gradually affected by stratification. The Ozmidov wavenumber exceeds the Kolmogorov wavenumber at $t \sim 7$, from which even the smallest size of velocity fluctuation is under the buoyancy effect. We will later discuss the spatial distribution of the scalar fluctuation and the energy spectra at $t = 4$, at which only large scale motions are affected by the buoyancy, and $t = 30$, at which a long time has passed since the buoyancy effect comes into play even for the smallest scale of eddies.

We note that the wavenumber, at which the Kolmogorov and Ozmidov wavenumbers agree, does not depend on the Prandtl number. This is because equating them leads to $\epsilon_* = \nu_* N_*^2$, and thus $k_{K*} = k_{O*} = (N_*/\nu_*)^{1/2}$ or $k_K = k_O = (Re_0/Fr_0)^{1/2}$ in the non-dimensional form. Since the initial Reynolds and Froude numbers are fixed in this study, the Kolmogorov wavenumber and the Ozmidov wavenumber agree at $k_K = k_O = (50/1)^{1/2} \sim 7.1$.

The spatial distribution of the potential energy $T^2/2Fr_0^2$ at $t = 4$ is shown in Fig. 3. It is immediately recognised that the scalar of $Pr = 700$ (salinity) contains smaller scale fluctuations than $Pr = 7$ (heat). This is simply because a scalar of higher Prandtl number has a smaller diffusion coefficient, so that small scale fluctuations can persist for a longer time. The distribution is almost isotropic for both $Pr = 7$ and $Pr = 700$ because the buoyancy effect is limited to relatively large scales at this time and anisotropy is not clearly visualised in these figures.

The spatial distributions of kinetic energy $u_i^2/2$ also exhibit isotropy and $Pr$-dependence, although the figures are not shown here. The isosurfaces of the kinetic energy for $Pr = 700$ have small wrinkles which are not observed in the case of $Pr = 7$, suggesting that the small-scale fluctuations of velocity is enhanced by the

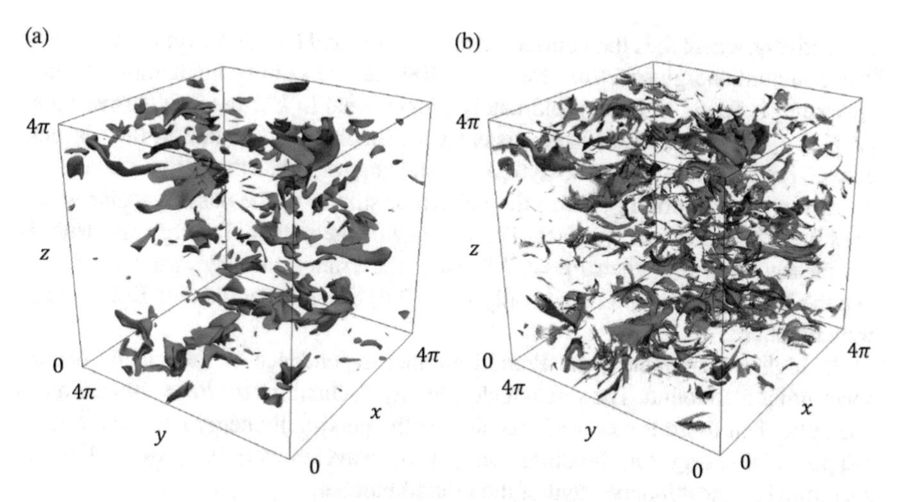


**Fig. 3** Spatial distribution of the potential energy $T^2/2Fr_0^2$ at $t = 4$ for **(a)** $Pr = 7$ and **(b)** $Pr = 700$. Isosurfaces of 25% of the maximum value are depicted

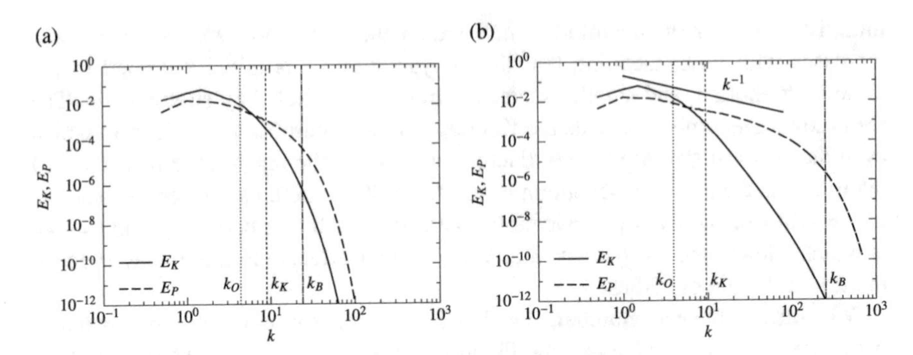


**Fig. 4** The kinetic energy spectrum $E_K$ and the potential energy spectrum $E_P$ at $t = 4$ for (**a**) $Pr = 7$ and (**b**) $Pr = 700$. The *vertical lines* show the Ozmidov wavenumber $k_O$, the Kolmogorov wavenumber $k_K$ and the Batchelor wavenumber $k_B$

energy conversion from the potential energy through the intermediary of the vertical scalar flux.

Figure 4 presents the kinetic and potential energy spectra at $t = 4$. They are defined by

$$E_K(k) = \int_{|\boldsymbol{k}|=k} \frac{1}{2}|\hat{u}_i(\boldsymbol{k})|^2 d\boldsymbol{k}, \tag{4}$$

and

$$E_P(k) = \int_{|\boldsymbol{k}|=k} \frac{1}{2Fr_0^2}|\hat{T}(\boldsymbol{k})|^2 d\boldsymbol{k}, \tag{5}$$

respectively, where $\hat{u}_i$ is the Fourier coefficient of $u_i$ and $\boldsymbol{k}$ is the wavenumber vector. The potential energy spectrum for $Pr = 700$, i.e. the salinity fluctuation variance spectrum, is found to be approximately proportional to $k^{-1}$ in an extensive range ($2 \lesssim k \lesssim 50$). Our result is consistent with Batchelor's $k^{-1}$ law in the viscous-convective subrange ($k_K \lesssim k \lesssim k_B$) [1] and suggests that the active scalar such as salinity could initially exhibit the behaviour similar to the passive scalar which does not have a buoyancy effect. The $k^{-1}$ law of the potential energy spectrum is not clearly observed when $Pr = 7$ because the Prandtl number is not very large and the Batchelor wavenumber is only $\sqrt{7}(\sim 2.6)$ times larger than the Kolmogorov wavenumber.

At high wavenumbers, the Prandtl-number dependence of the kinetic energy spectrum is also found. The small-scale velocity fluctuations for $Pr = 700$ are more energetic than those for $Pr = 7$ because of the persistent energy conversion from the potential energy. On the other hand, at low wavenumbers ($k \lesssim k_K \simeq 10$), the spectrum is almost independent of the Prandtl number.

We next show in Fig. 5 the spatial distribution of potential energy at $t = 30$, i.e. when the Ozmidov wavenumber is one order of magnitude larger than

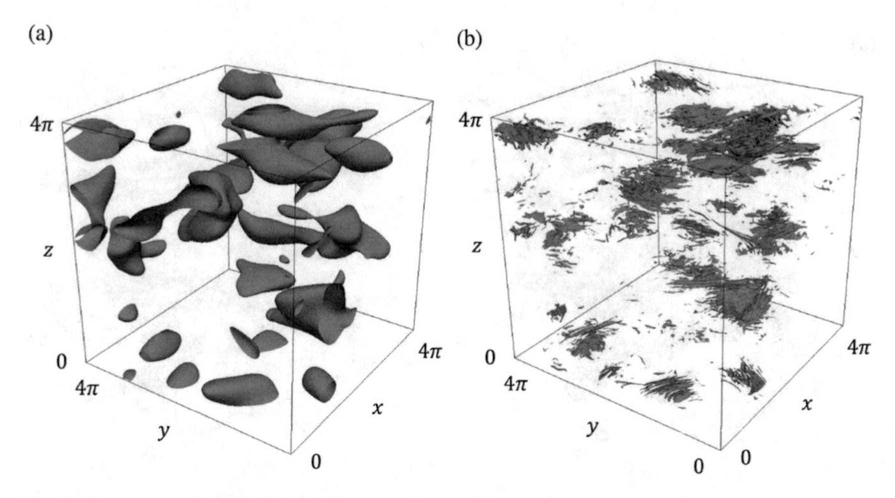


**Fig. 5** Spatial distribution of the potential energy $T^2/2Fr_0^2$ at $t = 30$ for (**a**) $Pr = 7$ and (**b**) $Pr = 700$. Isosurfaces of 30% of the maximum value are depicted

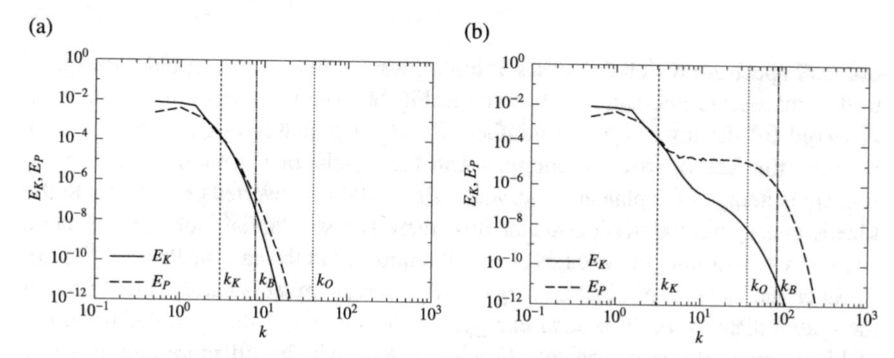


**Fig. 6** The kinetic energy spectrum $E_K$ and the potential energy spectrum $E_P$ at $t = 30$ for (**a**) $Pr = 7$ and (**b**) $Pr = 700$. The *vertical lines* show the Ozmidov wavenumber $k_O$, the Kolmogorov wavenumber $k_K$ and the Batchelor wavenumber $k_B$

the Kolmogorov wavenumber, and even the smallest eddies are strongly affected by buoyancy. The large-scale structures containing much of the potential energy (grey blobs) are the horizontally flat "pancake" structures typical to the stratified turbulence (e.g. [6]). These large-scale structures are independent of $Pr$, and similar in Fig. 5a, b. A significant difference between the two figures is that small scale fluctuations exist only in the blobs of $Pr = 700$, and they show an intermittent pattern. It can be viewed as a spatial localisation of scalar fluctuations.

The kinetic and potential energy spectra at the same time ($t = 30$) are presented in Fig. 6. The potential energy spectrum for $Pr = 700$ (Fig. 6b) no longer shows the $k^{-1}$ power law observed earlier ($t = 4$ in Fig. 4b), and a flat spectrum or a plateau ($\propto k^0$) appears between the Kolmogorov wavenumber and the Batchelor wavenumber. The plateau would contain much larger energy compared to the

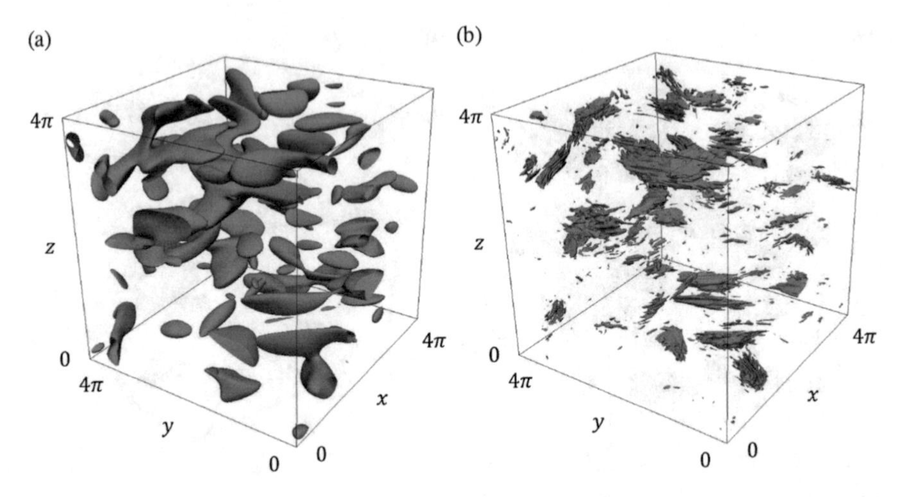


**Fig. 7** Spatial distribution of the kinetic energy dissipation rate $(\partial u_i/\partial x_j)^2/Re_0$ at $t = 30$ for (**a**) $Pr = 7$ and (**b**) $Pr = 700$. Isosurfaces of 30% of the maximum value are depicted

ordinary spectrum which decreases at higher wavenumbers, and would correspond to the small-scale fluctuations observed in Fig. 5b. The flat spectrum has not been observed for the lower Prandtl number ($Pr = 7$) probably due to the proximity between the Komogorov wavenumber and the Batchelor wavenumber. Since the potential energy in the plateau observed at $Pr = 700$ is converted persistently to the kinetic energy via the vertical scalar flux, there is a substantial increase of kinetic energy in sub-Kolmogorov scales ($k \sim 20$) compared to the case of $Pr = 7$ (Fig. 6).

Since the increase of kinetic energy at high wavenumber is hardly discernible in the spatial distribution of kinetic energy, we show in Fig. 7 the spatial distribution of kinetic energy dissipation rate $(\partial u_i/\partial x_j)^2/Re_0$, which will more clearly show the small-scale structures. Small-scale fluctuations of the kinetic-energy dissipation rate and an intermittent pattern are found when $Pr = 700$ (Fig. 7b), similar to the potential energy distribution observed in Fig. 5b. The maximum value of the local kinetic energy dissipation rate at $Pr = 700$ is about 70% larger than that at $Pr = 7$, because of the increase of the kinetic energy in the sub-Kolmogorov scale. We note that the kinetic energy dissipation spectrum also has a small plateau around $k \sim 20$ though the figure is not shown here.

## 4 Conclusions

We have demonstrated a direct numerical simulation of turbulence in a fluid stratified by a high Prandtl-number scalar such as heat ($Pr = 7$) or salinity ($Pr = 700$), using $4096^3$ grid points. Massive computation has been performed using the 1024 nodes of NEC SX-ACE, which have enabled us to resolve the smallest scale of salinity fluctuations about 26 times smaller than the Kolmogorov scale. The results can be summarised as follows.

Initially, before the Ozmidov wavenumber exceeds the Kolmogorov wavenumber, buoyancy effect is limited to the scales larger than the Ozmidov scale. Then, the small-scale scalar fluctuations exist isotropically in space, and the potential energy for $Pr = 700$ has a spectrum proportional to $k^{-1}$ at high wavenumbers between the Kolmogorov wavenumber and the Batchelor wavenumber (viscous-convective subrange). These results show that the initial behaviour of an active (buoyant) scalar is very similar to the passive scalar.

Later, after the Ozmidov wavenumber exceeds the Kolmogorov wavenumber, the buoyancy affects the fluid motion down to the Kolmogorov scale, which is the smallest scale of the fluid motion. In that final period of decay, the scalar fluctuations at $Pr = 700$ shows a spatially localised distribution, and its spectrum exhibits a plateau ($\propto k^0$) between the Kolmogorov and Batchelor wavenumbers. The potential energy in the wavenumber range of plateau is converted persistently to the kinetic energy via the vertical scalar flux, leading to an increase of kinetic energy (and its dissipation rate) at scales smaller than the Kolmogorov scale.

The localisation and the flat spectrum have not been observed at $Pr = 7$, possibly because the Kolmogorov wavenumber and the Batchelor wavenumber are not well separated. However, the reason why these two prominent features appear only at $Pr = 700$ is still an open question, and the energy budget of the spectrum is now under investigation in the hope of clarifying the mechanisms.

**Acknowledgements** This research used computational resources of the HPCI system provided by Tohoku University through the HPCI System Research Project (Project ID: hp160108), and the Earth Simulator at the Japan Agency for Marine Earth Science and Technology.

## References

1. Batchelor, G.K.: Small-scale variation of convected quantities like temperature in turbulent fluid. J. Fluid Mech. **5**, 113–133 (1959)
2. Gibson, C.H., Schwarz, W.H.: The universal equilibrium spectra of turbulent velocity and scalar fields. J. Fluid Mech. **16**, 365–384 (1963)
3. Bogucki, D., Domaradzki, J.A., Yeung, P.K.: Direct numerical simulations of passive scalar with $Pr > 1$ advected by turbulent flow. J. Fluid Mech. **343**, 111–130 (1997)
4. Yeung, P.K., Xu, S., Donzis, D.A., Sreenivasan, K.R.: Simulations of three-dimensional turbulence mixing for Schmidt numbers of the order 1000. Flow Turbul. Combust. **72**, 333–347 (2004)
5. Riley, J.J., Metcalfe, R.W., Weissman, M.A.: Direct numerical simulations of homogeneous turbulence in density-stratified fluids. In: B.J. West (ed.) Proceeding of the AIP Conference on Nonlinear Properties of Internal Waves, pp. 79–112 (1981)
6. Métais, O., Herring, J.R.: Numerical simulations of freely evolving turbulence in stably stratified fluid. J. Fluid Mech. **202**, 117–148 (1989)

# Wavelet-Based Compression of Volumetric CFD Data Sets

**Patrick Vogler and Ulrich Rist**

**Abstract** One of the major pitfalls of storing "raw" simulation results lies in the implicit and redundant manner in which it represents the flow physics. Thus transforming the large "raw" into compact feature- or structure-based data could help overcome the I/O bottleneck. Several compression techniques have already been proposed to tackle this problem. Yet, most of these so-called lossless compressors either solely consist of dictionary encoders, which merely act upon the statistical redundancies in the underlying binary data structure, or use a preceding predictor stage to decorrelate intrinsic spatial redundancies. Efforts have already been made to adapt image compression standards like the JPEG codec to floating-point arrays. However, most of these algorithms rely on the discrete cosine transform which offers inferior compression performance when compared to the discrete wavelet transform. We therefore demonstrate the viability of a wavelet-based compression scheme for large-scale numerical datasets.

## 1 Introduction

The steady increase of available computer resources has enabled engineers and scientists to use progressively more complex models to simulate a myriad of fluid flow problems. Yet, whereas modern high performance computers (HPC) have seen a steady growth in computing power, the same trend has not been mirrored by a significant gain in data transfer rates. Current systems are capable of producing and processing high amounts of data quickly, while the overall performance is oftentimes hampered by how fast a system can transfer and store the computed data. Considering that CFD researchers invariably seek to study simulations with increasingly higher temporal resolution on fine grained computational grids, the imminent move to exascale performance will consequently only exacerbate this problem [10].

P. Vogler (✉) • U. Rist
Institute of Aerodynamics and Gas Dynamics, University of Stuttgart, Pfaffenwaldring 21, 70569 Stuttgart, Germany
e-mail: patrick.vogler@iag.uni-stuttgart.de; rist@iag.uni-stuttgart.de


M.M. Resch et al. (eds.), *Sustained Simulation Performance 2017*,
DOI 10.1007/978-3-319-66896-3_8

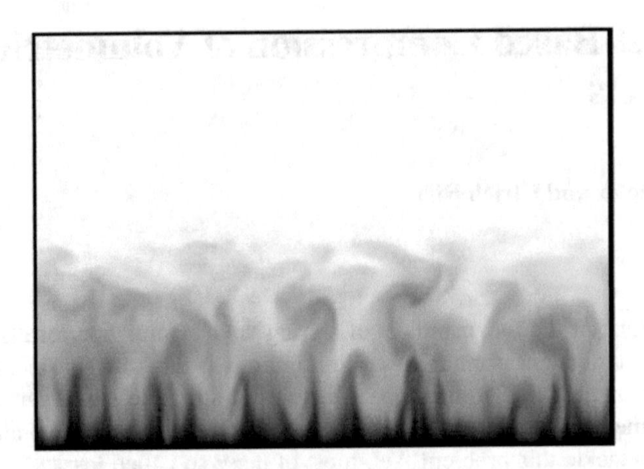

**Fig. 1** Stream-wise velocity field from a numerical simulation of a turbulent flat-plate boundary layer flow at $Re_{\theta,0} = 300$

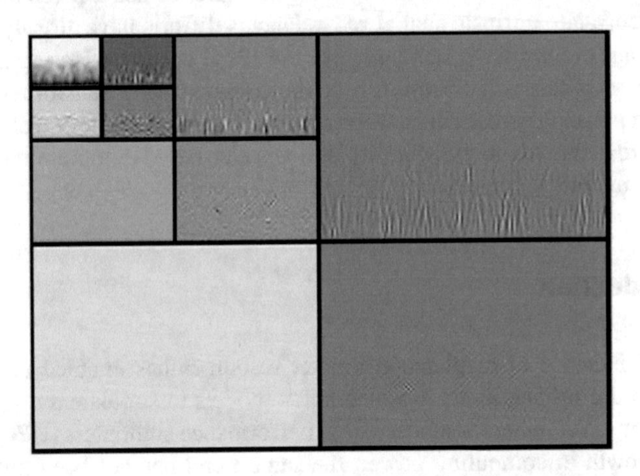

**Fig. 2** Dyadic decomposition into subbands for the streamwise velocity field of Fig. 1

One way to alleviate the I/O bottleneck would be to reduce the number of time steps which are written to the file system. While this trivial data reduction method may be tolerable for simulations that reach a steady state solution after a minuscule amount of time, the same approach would be fatal for highly transient physical phenomena. Considering that most fluid flow problems are subject to diffusion, however, we can conclude that our numerical datasets will typically be smooth and continuous, resulting in a frequency spectrum that is dominated by lower modes (see Figs. 1 and 2) [10]. Thus our best way forward should be to use the otherwise wasted compute cycles by exploiting these inherent statistical redundancies to create a more compact form of the information content. Since effective data storage is a pervasive problem in information technology, much effort has already been spent on developing newer and better compression algorithms. Most of the prevalent compression techniques that are available for floating-point datasets, however, are so-called lossless compressors that either solely consist of dictionary encoders, which merely act upon the statistical redundancies in the underlying binary data structure (i.e. Lempel-Ziv-Welch algorithm [13]), or use a preceding predictor stage to decorrelate intrinsic spatial redundancies (i.e. FPZIP [6]). These compression schemes are limited to a size reduction of only 10–30%, not allowing for a more efficient compression by neglecting parts of the original data that contribute little to the overall information content [7, 8].

Prominent compression standards that allow for lossy compression, however, can be found in the world of entertainment technology. In this context, Loddoch and Schmalzl [8] have extended the Joint Photographic Experts Group (JPEG) standard for volumetric floating-point arrays by applying the one-dimensional real-to-real discrete cosine transform (DCT) along the axis of each spatial dimension. The distortion is then controlled by a frequency dependent damping of the DCT coefficients followed by a quantization stage which maps the floating-point values onto an integer range with adjustable size. Finally, the quantized DCT coefficients are encoded using a variable-length code similar to that proposed by the JPEG standard. Lindstrom [7], on the other hand, uses the fixed point number format Q, which maps the floating point values onto the dynamic range of a specified integer type, followed by a lifting based integer-to-integer implementation of the discrete cosine transform. The DCT coefficients are then encoded using an embedded coding algorithm to produce a quality-scalable codestream. While these compression algorithms are simple and efficient in exploiting the low frequency nature of most numerical datasets, their major disadvantage lies in the non-locality of the basis functions of the discrete cosine transform. Thus, if a DCT coefficient is quantized, the effect of a lossy compression stage will be felt throughout the entire flow field [3]. To alleviate this, the numerical field is typically divided into small blocks and the discrete cosine transform is applied to each block one at a time. While partitioning the flow field also facilitates random-access read and write operations, this approach gives rise to block boundary artifacts which are synonymous with the JPEG compression standard.

In order to circumvent this problem, we propose to adapt the JPEG-2000 (JP2) compression standard for volumetric floating-point arrays. In contrast to the baseline JPEG standard, JPEG-2000 employs a lifting-based one-dimensional discrete wavelet transformed (DWT) that can be performed by either the reversible LeGall-(5,3) taps filter for lossless or the non reversible Daubechies-(9,7) tabs filter for lossy coding [1]. Due to its time-frequency representation, which identifies the time or location at which various frequencies are present in the original signal, the discrete wavelet transform allows for the full frame decorrelation of large scale numerical datasets. This eliminates blocking artifacts at high compression ratios, commonly associated with the JPEG standard. Furthermore, the dyadic decomposition into multi-resolution subbands (see Fig. 2) enables the compression standard to assemble a resolution-scalable codestream and the definition of so-called Regions of Interest (ROI), which are to be coded and transmitted with better quality

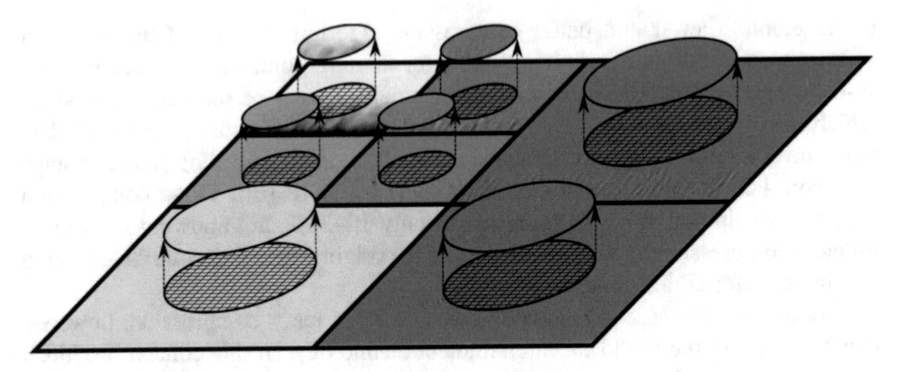

Fig. 3 ROI mask generation in the wavelet domain

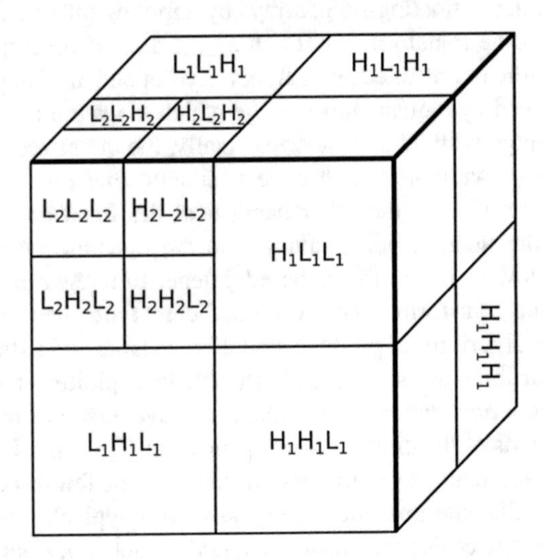


Fig. 4 3-Dimensional dyadic decomposition into subbands [2]

and less distortion than the rest of the flow field (see Fig. 3). Similar to the ZFP Standard by Lindstrom [7], JPEG-2000 employs an embedded coding algorithm to produce a quality scalable codestream [12]. Finally, JPEG-2000s volumetric extension (JP3D) translates these same capabilities to multi-dimensional datasets by applying the one-dimensional discrete wavelet transform along the axis of each subsequent dimension (see Fig. 4) [2].

In this paper we will assess the lossy compression stage of the JPEG-2000 codec for volumetric floating-point arrays and how it compares to a DCT (ZFP) and dictionary encoder (7-Zip) based compression scheme. In Sect. 2 we will provide an overview of the fundamental building blocks of the JPEG-2000 Part 1 codec and its volumetric extension JP3D. Our focus will lie on the transformation and quantization stage, since most of our efforts to adapt the JPEG-2000 algorithm to floating-point numbers have been spent on this part of the codec. For a thorough discussion of the Embedded Block Coding with Optimized Truncation (EBCOT) algorithm the reader is referred to the treatise by Taubman and Marcellin [12]. Preliminary compression results for a numerical simulation of a turbulent flat-plate boundary layer flow and Taylor-Green vortex decay will be shown in Sect. 3. Finally, we conclude with some closing remarks in Sect. 4.

## 2 Technical Description

The JPEG-2000 standard is divided into 14 parts, with part 1 defining the core coding system of the compression standard. Each subsequent part, such as the volumetric extension JP3D, translates the capabilities of the baseline codec to different fields of application [12]. Since we are, for the most part, only interested in compressing three-dimensional numerical datasets, we will limit our discussion to the baseline codec and its volumetric extension.

As depicted in Fig. 5, the first step in our compression stage is to generate a time-frequency representation of the original data samples, which enables relatively simple quantization and coding operations. It is worth noting that, given an invertible Transform $T$, this step will not introduce any distortion in the decompressed dataset [12]. In the second step, the transformed samples are represented using a sequence of quantization indices. This mapping operation introduces distortion in our decompressed data, since the set of possible outcomes for each quantization index is generally smaller than for the transformed samples [1]. Finally, the quantization indices are losslessly entropy coded to form the final bit-stream [9]. To adapt the JPEG-2000 codec for floating point arrays, we use the fixed point number format Q as described in Sect. 2.1. Next, the transform (Sect. 2.2) and quantization (Sect. 2.3) components are discussed in more detail.

### 2.1 Fixed Point Number Format

Our approach to handle floating point values was inspired by the ZFP compression standard [7]. Using the fixed point number format Q, we first map the floating point values onto the dynamic range of a specified integer type (i.e. 64 bit). To this end,

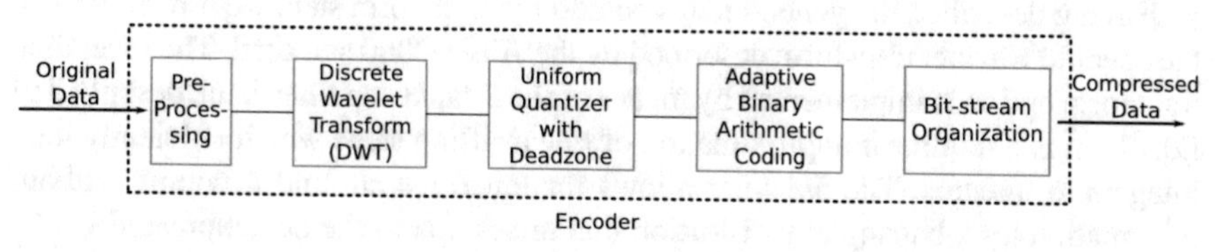


**Fig. 5** Fundamental building blocks of the JPEG-2000 compression stage [9]

the conservative variables $\rho$, $\rho u$, $\rho v$, $\rho w$ and $E$ were first aligned with regards to their largest floating point exponents, which are stored uncompressed in the header of the codestream. We used a Q8.23 two's compliment format that allows numbers in the range $[-255, 255)$ to be represented. Although the normalized floating point values lie in a smaller range $(-1, +1)$, the additional dynamic range was added to prevent an overflow of the variables during the subsequent transform stage.

## 2.2 *The Discrete Wavelet Transform*

The transform is responsible for massaging the data samples into a more amenable representation for compression. It should capture the statistical dependencies among the original samples and separate relevant from irrelevant information for an optimal quantization stage. Finally, the use of integer DWT filters allows for both lossy and lossless compression in a single code-stream [9].

The lifting-based one-dimensional discrete wavelet transform is best understood as a pair of low- and high-pass filters, commonly known as the analysis filter-bank. Successive application of the analysis filter pair is followed by a down-sampling operation by a factor of two, discarding odd indexed samples. The analysis filter-bank is designed in such a manner that perfect reconstruction, barring any quantization error, is still possible after the downsampling step. The low pass filter attenuates high frequency components in a one-dimensional signal, resulting in a blurred version of the original dataset. This low-pass output is typically highly correlated and thus can be subjected to further stages of the analysis filter-bank. The high-pass filter, on the other hand, preservers the high frequency components, which usually results in a sufficiently decorrelated high-pass signal. Consequently, most DWT decompositions only further decompose the low-pass output to produce what is known as a dyadic decomposition [9].

The filtered samples, which are output from the transform operation, are referred to as wavelet coefficients. To ensure the efficiency of the compression algorithm, these coefficients are critically sampled by virtue of the downsampling operation. This means, that the total number of wavelet coefficients needs to be equal to the number of original signal samples. Thus, when the DWT decomposition is applied to an odd-length signal, either the low- or high-pass sequence will have one additional sample. This choice is dictated by the position of the odd-length signal in relation to the global coordinate system [12].

Having described the general ideas behind the transform step, we now introduce the specific wavelet transform described by the JPEG-2000 standard. The reversible transform option is implemented by means of the 5-tap/3-tap filter-bank described in Eq. (1). It is a nonlinear approximation of linear lifting steps which efficiently map integers to integers. The 5/3 filter allows for repetitive en- and decoding without information loss, barring any distortion that arises due to the decompressed image

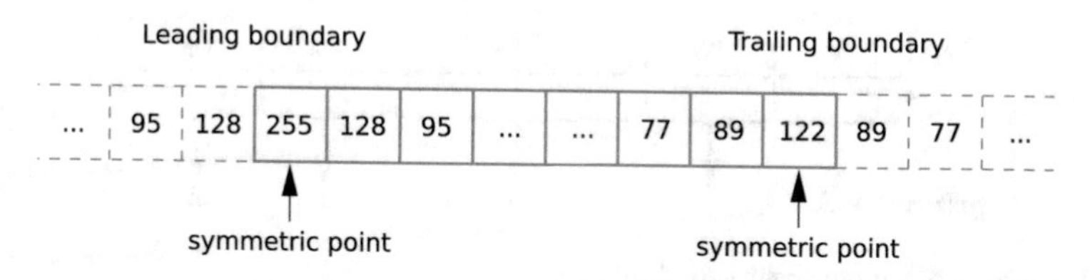


**Fig. 6** Symmetric extension at the leading and trailing boundaries of a signal segment

values being clipped, should they fall outside their full dynamic range [12].

$$
\begin{aligned}
y(2n+1) &= x(2n+1) - \left\lfloor \frac{x(2n)+x(2n+2)}{2} \right\rfloor, \\
y(2n) &= x(2n) - \left\lfloor \frac{x(2n-1)+x(2n+1)+2}{4} \right\rfloor.
\end{aligned} \tag{1}
$$

While the 5/3 bi-orthogonal filter-bank is a prime example for a reversible integer-to-integer transform, its energy compaction, due to its nonlinearity, usually falls short of most floating point filter-banks. The most prominent real-to-real transform is the irreversible 9-tap/7-tap filter-bank described in Eq. (2) [9].

$$
\begin{aligned}
y(2n+1) &\leftarrow x(2n+1) + (-1.586 \times \lfloor x(2n) + x(2n+2) \rfloor), \\
y(2n) &\leftarrow x(2n) + (-0.052 \times \lfloor y(2n-1) + y(2n+1) \rfloor), \\
y(2n+1) &\leftarrow y(2n+1) + (0.883 \times \lfloor y(2n) + y(2n+2) \rfloor), \\
y(2n) &\leftarrow y(2n) + (0.443 \times \lfloor y(2n-1) + y(2n+1) \rfloor), \\
y(2n+1) &\leftarrow -1.230 \times y(2n+1), \\
y(2n) &\leftarrow (1/1.230) \times y(2n).
\end{aligned} \tag{2}
$$

To ensure the perfect reconstruction property of the wavelet transform, the undefined samples outside of the finite-length signal segment need to be filled with values related to the samples inside the signal segment. When using odd-tap filters, the signal is symmetrically and periodically extended as shown in Fig. 6 [5].

## 2.3 *Quantization*

Unlike its predecessor, the JPEG-2000 algorithm employs a central deadzone quantizer (see Fig. 7) to reduce the inherent entropy in the wavelet coefficients. This reduction in precision is lossy, unless the quantization stepsize is set to 1 and the subband samples are integers. Each of the wavelet coefficients $y_b$ of the subband $b$

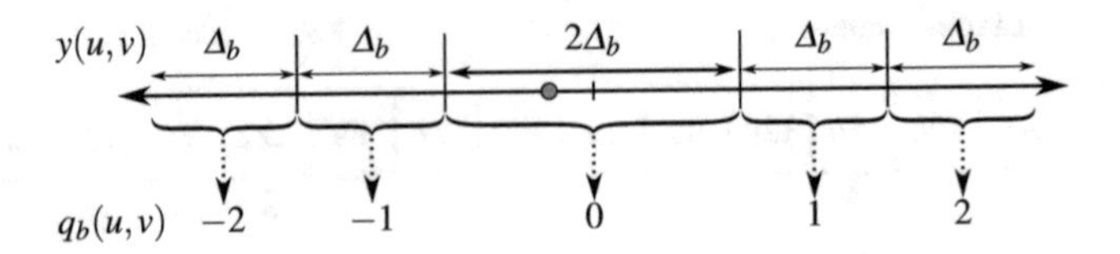


**Fig. 7** Uniform scalar quantizer with deadzone

is mapped to the quantization value $q_b$ according to the formula [12]:

$$q_b = sign\,(y_b\,[\mathbf{n}]) \left\lfloor \frac{|y_b\,[\mathbf{n}]|}{\Delta_b} \right\rfloor. \tag{3}$$

the stepsize $\Delta_b$ is calculated as follows:

$$\Delta_b = 2^{R_b - \epsilon_b} \left(1 + \frac{\mu_b}{2^{26}}\right), \tag{4}$$

$$0 \leq \epsilon_b < 2^6, \quad 0 \leq \mu_b < 2^{26}. \tag{5}$$

where $\epsilon_b$ is the exponent, $\mu_b$ is the mantissa of the corresponding stepsize and $R_b$ represents the dynamic range of the subband $b$. This limits the largest possible stepsize to twice the dynamic range of the subband. In case of a reversible coding path, $\Delta_b$ is set to one by choosing $\mu_b = 0$ and $R_b = \epsilon_b$ [12].

# 3 Experimental Results

## 3.1 Data Sets

For testing the JPEG-2000 algorithm we used datasets from numerical simulations of a turbulent flat-plate boundary layer flow at $Ma_\infty = 0.3$ and $Ma_\infty = 2.5$ (see Fig. 8) performed by C. Wenzel at the Institute of Aerodynamics [14]. The spatial resolution of the numerical grid was set to $nx \times ny \times nz = 3300 \times 240 \times 512$ nodes in streamwise, wall-normal and spanwise directions respectively. The file size for one time step containing the conservative variables $\rho$, $\rho u$, $\rho v$, $\rho w$, $E$ amounted to 16.220.192.238 bytes.

To compare the JPEG-2000 algorithm with the ZFP and 7-Zip encoders, we used a numerical simulation of a Taylor-Green vortex decay at $Re = 1600$, with a spatial resolution of $nx \times ny \times nz = 256 \times 256 \times 256$ (see Fig. 9). The compression performance was evaluated at the non-dimensional time-steps $t = 0, 2.5, 5, 7.5, 10, 12.5, 15, 17.5, 20$. The file size for one time step containing the conservative variables $\rho$, $\rho u$, $\rho v$, $\rho w$ and $\rho E$ amounted to 703.052.304 bytes.

To assess the overall quality of the decompressed file we used the peak signal-to-noise ratio metric (PSNR), which is evaluated based on the mean-square-error

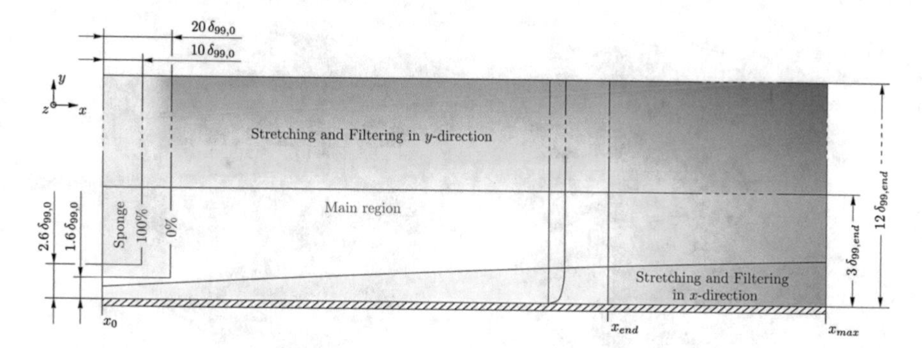


**Fig. 8** Numerical setup for the simulation of a turbulent flat-plate boundary layer flow at $Ma_\infty = 0.3$ and $Ma_\infty = 2.5$ [14]

(MSE):

$$MSE = \frac{1}{ijk} \sum_{x=1}^{i} \sum_{y=1}^{j} \sum_{z=1}^{k} \left| I(x,y,z) - I'(x,y,z) \right|^2 , \tag{6}$$

$$PSNR = 20 \cdot \log_{10} \left( \frac{\max(I(x,y,z)) - \min(I(x,y,z))}{\sqrt{MSE}} \right) . \tag{7}$$

where $I(x, y, z)$ is the original, $I'(x, y, z)$ the decompressed data and $i, j, k$ the dimensions of the volumetric dataset. The PSNR is expressed in dB (decibels). We found that good reconstructed datasets typically have PSNR values of 90 dB or more. We ran our experiments on a single core of an Intel Core i7-6700 processor with 3.40 GHz and 32 GB of 2133 MHz DDR4 RAM.

## 3.2 Results

Figure 10 shows a close-up of the original (top) and compressed (bottom) DNS for a turbulent flat-plate flow at $Ma_\infty = 0.3$, Fig. 11 for a turbulent flat-plate flow at $Ma_\infty = 2.5$. The compression ratio for the simulation at $Ma_\infty = 0.3$ measured 18:1 with a PSNR of 171.1, while the compression ratio for the simulation at $Ma_\infty = 2.5$ measured 16:1 with a PSNR of 171.0. The average compression time amounted to 936 s. In comparison, the compression ratio for the lossless LZMA Algorithm (7-Zip) measured only 1.2–1.3:1, with an average compression time of 1743 s.

Overall the JPEG-2000 algorithm offers good compression ratios with reasonably well reconstruction of the numerical datasets. Due to the floating point arithmetic and the many-to-one mapping, however, information will be irreversibly lost during the preprocessing of the data samples and thus true lossless compression cannot be achieved. Furthermore, the entropy encoder is unable to take full advantage of its optimal truncation algorithm since a large part of the quantization

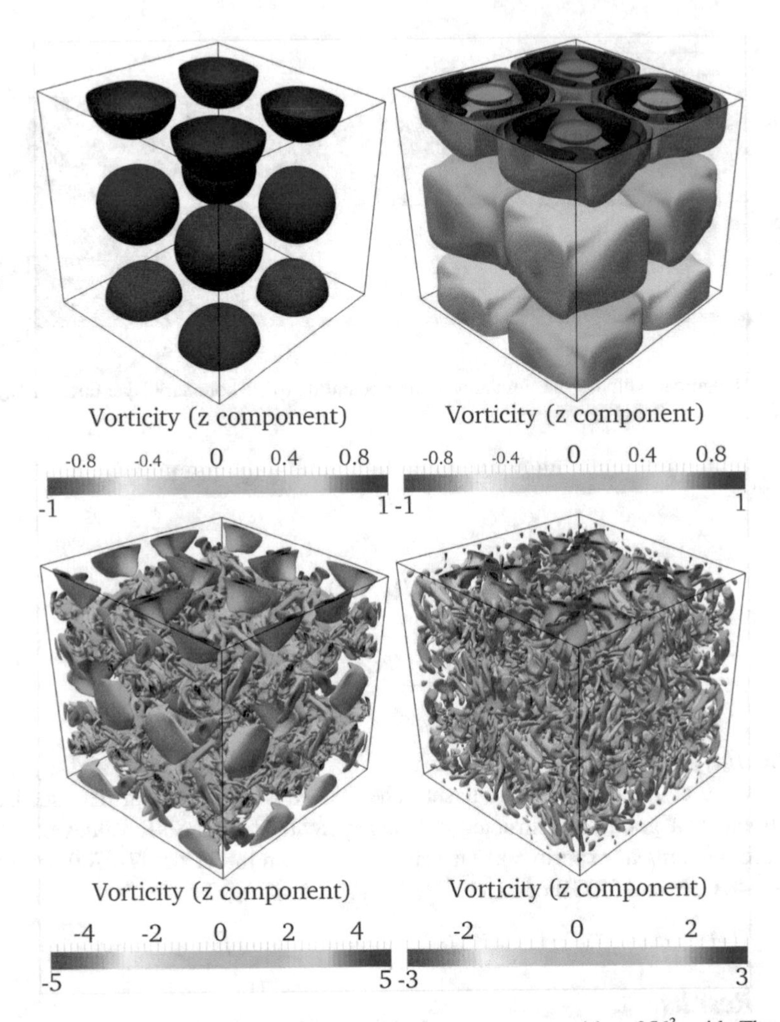


**Fig. 9** Visualization of the Taylor-Green vortex decay test case with a $256^3$ grid. The non-dimensional vorticity (z-component) is shown from *top left to bottom right* for the non-dimensional times t = 0, 2.5, 10, 20 [4]

(fixed point number transformation) falls outside of its purview. Furthermore, it is worth noting that the processor time spent for compressing the data sets is excessively high. This, however, can be attributed to the fact that the JPEG-2000 Codec used in this paper has not yet been optimized for speed.

Table 1 compares the compression ratio, compression time and PSNR for the JP2, ZFP and 7-Zip compressors. Several non-dimensional time-steps for the numerical simulation of a Taylor-Green vortex decay are listed to assess the effects of different vortex scales on the overall compression performance. We found that both the ZFP and JP2 offer significantly larger compression ratios when compared to the

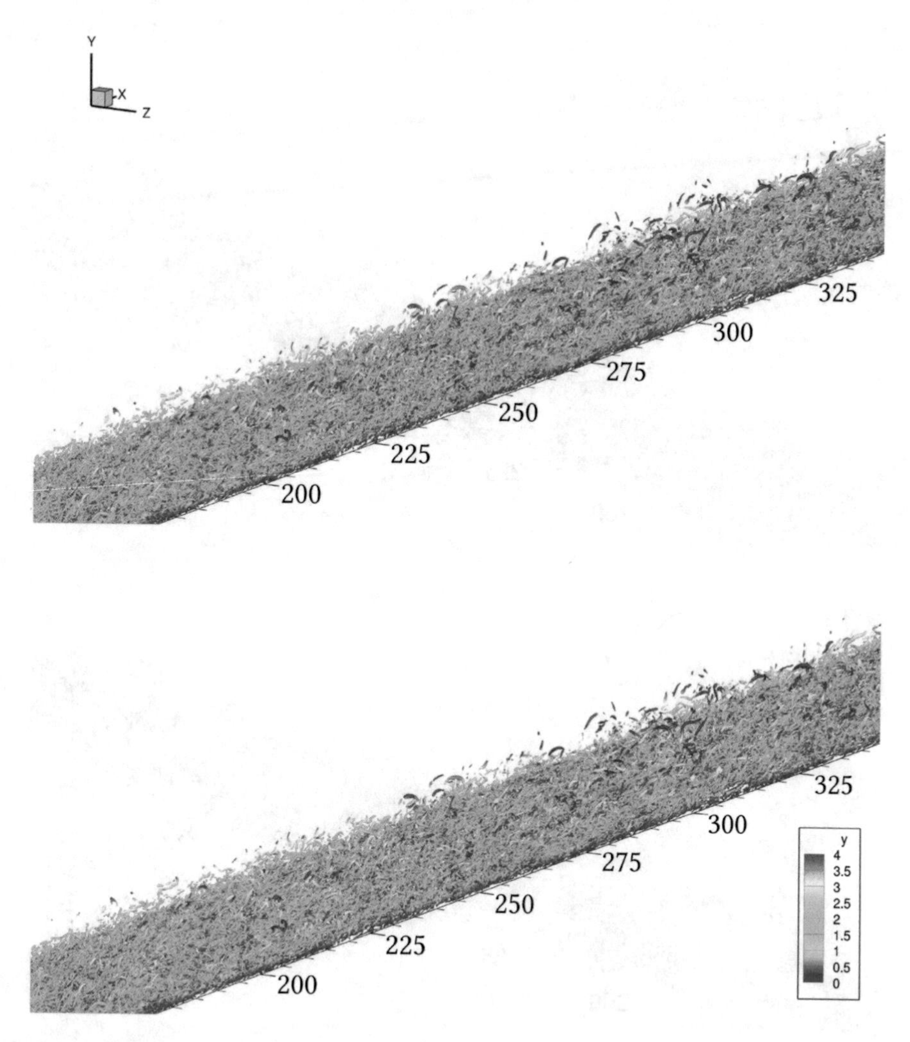


**Fig. 10** Original (*top*) and Compressed (*bottom*) DNS of a turbulent flat-plate boundary layer flow at $Ma_\infty = 0.3$. Flow structures identified by the $\lambda_2$-criterion for $\lambda_2 = -0.15$. Coloration of isosurfaces according to wall normal distance $y$ [14]

7-Zip encoder. The overall distortion of the flow field was observed to be small for both lossy algorithms, with the JP2 compressor offering a considerably better reconstruction of the numerical dataset at similar compression ratios. In contrast, the compression time for our wavelet-based approach is substantially larger. It is, however, noteworthy that most of the processing time is spent on the entropy encoding stage of the JPEG-2000 compressor. Since both the ZFP and JPEG-2000 Codec share a similar embedded coding algorithm we feel confident, that this large discrepancy in compression time is due to the unoptimized nature of our compression algorithm.

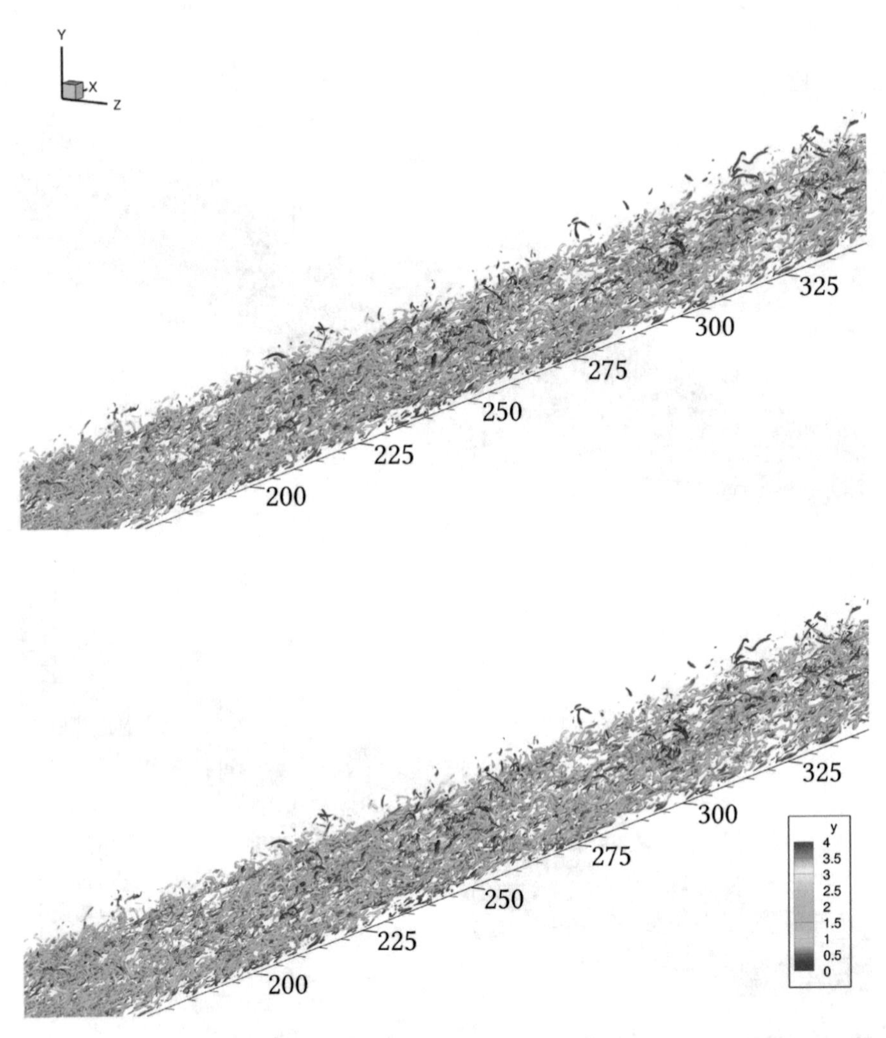


**Fig. 11** Close-up of the Original (*top*) and Compressed (*bottom*) DNS of a turbulent flat-plate boundary layer flow at $Ma_\infty = 2.5$. Flow structures identified by the $\lambda_2$-criterion for $\lambda_2 = -0.15$. Coloration of isosurface according the wall normal distance $y$ [14]

# 4 Conclusions

We presented a wavelet-based lossy compression scheme that allows for the compression of large volumetric floating-point arrays. The proposed technique is based on the JPEG-2000 algorithm [12]. A comparison with established compression techniques was done using three-dimensional data sets from a numerical simulation of a turbulent flat-plate boundary layer flow and Taylor-Green vortex decay. Based on the results of our study we found that our compression approach can significantly

**Table 1** Compression ratio, compression time and peak signal to noise ratio for the JP3D, ZFP and 7-Zip compressor

| | Compression ratio | | | Compression time | | | PSNR | | |
|---|---|---|---|---|---|---|---|---|---|
| Timestep | JP2 | ZFP | 7-Zip | JP2 | ZFP | 7-Zip | JP2 | ZFP | 7-Zip[a] |
| 0 | 60.90 | **61.10** | 41.72 | 20.10 | **1.61** | 114.91 | **181.90** | 64.15 | ∞ |
| 2954 | 29.45 | **29.50** | 6.03 | 32.37 | **2.43** | 160.37 | **181.30** | 86.98 | ∞ |
| 5909 | 13.58 | **13.60** | 5.84 | 66.38 | **2.85** | 173.20 | **182.90** | 75.98 | ∞ |
| 8864 | **8.23** | 8.22 | 5.57 | 94.78 | **3.06** | 164.00 | **182.70** | 86.23 | ∞ |
| 11819 | **6.53** | **6.53** | 5.07 | 109.19 | **3.17** | 172.91 | **178.20** | 91.36 | ∞ |
| 14774 | **6.28** | **6.28** | 4.63 | 118.34 | **3.29** | 172.80 | **179.20** | 98.42 | ∞ |
| 17729 | **6.21** | **6.21** | 4.21 | 116.37 | **3.34** | 186.38 | **175.60** | 100.3 | ∞ |
| 20684 | **6.36** | **6.36** | 3.92 | 115.25 | **3.39** | 183.88 | **175.80** | 102.4 | ∞ |
| 23634 | **6.47** | 6.46 | 3.72 | 113.85 | **3.34** | 181.45 | **175.3** | 105.5 | ∞ |

The best results appear in bold

[a]The PSNR for the 7-Zip compressor is always infinity due to its lossless nature

decrease the overall flow field distortion while maintaining a high compression ratio. The results, however, also show the need for an optimized implementation of our wavelet based codec. Considering our compressor shares a similar entropy encoding stage with the ZFP codec, we anticipate that improvements to our algorithms will lead to competitive compression and decompression times.

Looking ahead, our hope is that so-called intraband prediction methods, which are used in the High Efficiency Video Coding standard (HEVC), could further increase the overall efficiency [11, 15]. We will also investigate higher order signal transforms which allow for the efficient transformation of images with smooth regions separated by smooth boundaries.

**Acknowledgements** This work was supported by a European Commission Horizon 2020 project grant entitled "ExaFLOW: Enabling Exascale Fluid Dynamics Simulations" (grant reference 671571). The authors would also like to thank Christoph Wenzel at the Institute of Aerodynamics and Gasdynamics at the University of Stuttgart for providing labor intensive data sets of a turbulent flat-plate boundary flow.

## References

1. Acharya, T., Tsai, P.: JPEG2000 Standard for Image Compression: Concepts, Algorithms and VLSI Architectures. Wiley, New Jersey (2005)
2. Bruylants, T., Munteanu, A., Schelkens, P.: Wavelet based volumetric medical image compression. Signal Process. Image Commun. **31**, 112–133 (2015). doi:10.1016/j.image.2014.12.007
3. Christopoulos, C., Skodras, A., Ebrahimi, T.: The JPEG2000 still image coding system: an overview. IEEE Trans. Consum. Electron. **46**(4), 1103–1127 (2000). doi:10.1109/30.920468
4. Jacobs, T., Jammy, S.P., Sandham, N.D.: OpenSBLI: a framework for the automated derivation and parallel execution of finite difference solvers on a range of computer architectures. J. Comput. Sci. **18**, 12–23 (2017). doi:10.1016/j.jocs.2016.11.001

5. Li, S., Li, W.: Shape-adaptive discrete wavelet transforms for arbitrarily shaped visual object coding. IEEE Trans. Circuits Syst. Video Technol. **10**(5), 725–743 (2000). doi:10.1109/76.856450
6. Lindstrom, P.: Fixed-rate compressed floating-point arrays. IEEE Trans. Vis. Comput. Graph. **20**(12), 2674–2683 (2014). doi:10.1109/TVCG.2014.2346458
7. Lindstrom, P., Isenburg M.: Fast and efficient compression of floating-point data. IEEE Trans. Vis. Comput. Graph. **12**(5), 1245–1250 (2006). doi:10.1109/TVCG.2006.143
8. Loddoch, A., Schmalzl, J.: Variable quality compression of fluid dynamical data sets using a 3-D DCT technique. Geochem. Geophys. Geosyst. **7**(1), 1–13 (2006). doi:10.1029/2005GC001017
9. Rabbani, M., Joshi, R.: An overview of the JPEG 2000 still image compression standard. Signal Process. Image Commun. **17**(1), 3–48 (2002). doi:10.1016/S0923-5965(01)00024-8
10. Schmalzl, J.: Using standard image compression algorithms to store data from computational fluid dynamics. Comput. Geosci. **29**, 1021–1031 (2003). doi:10.1016/S0098-3004(03)00098-0
11. Sze, V., Budagavi, M., Sullivan, G.J. (eds.): High Efficiency Video Coding (HEVC): Algorithms and Architectures. Springer, Cham (2014)
12. Taubman, D., Marcellin, M.: JPEG2000: Image Compression Fundamentals, Standards and Practice. Springer, New York (2002)
13. Welch, T.A.: A technique for high-performance data compression. Computer **17**(6), 8–19 (1984). doi:10.1109/MC.1984.1659158
14. Wenzel, C., Selent, B., Kloker, M., Rist, U.: DNS of compressible turbulent boundary layers and assessment of data-scaling-law quality. Under consideration for publication in J. Fluid Mech.
15. Wien, M.: High Efficiency Video Coding (HEVC): Coding Tools and Specification. Springer, Heidelberg (2014)

# Validation of Particle-Laden Large-Eddy Simulation Using HPC Systems

**Konstantin Fröhlich, Lennart Schneiders, Matthias Meinke, and Wolfgang Schröder**

**Abstract** In this contribution, results of a direct particle-fluid simulation (DPFS) are compared with direct numerical simulations and large-eddy simulations (LES) using a popular Euler-Lagrange method (ELM). DPFS facilitates the computation of particulate turbulent flow with particle sizes on the order of the smallest flow scales, which requires advanced numerical methods and parallelization strategies accompanied by considerable computing resources. After recapitulating methods required for DPFS, a setup is proposed where DPFS is used as a benchmark for direct numerical simulations and LES. Therefore, a modified implicit LES scheme is proposed, which shows convincing statistics in comparison to a direct numerical simulation of a single phase flow. Preliminary results of particle-laden flow show good agreement of the LES and the DPFS findings. Further benchmark cases for an appreciable range of parameters are required to draw a rigorous conclusion of the accuracy of the ELM.

## 1 Introduction

Particle-laden turbulent flow is of importance in a broad field of applications including natural and technical environments. Examples may be found in the settling of aerosol particles in atmospheric flows, in the transport of dust through the human respiration system, in fuel injections of internal combustion engines, as well as in the combustion of pulverized coal particles in a furnace. However, for particles with diameter $d_p \approx \eta_k$, with $\eta_k$ the Kolmogorov scale, there is no accurate and robust model available [1]. This may be explained by the numerous scales

K. Fröhlich (✉) • L. Schneiders • M. Meinke
Institute of Aerodynamics, RWTH Aachen University, Aachen, Germany
e-mail: k.froehlich@aia.rwth-aachen.de; l.schneiders@aia.rwth-aachen.de; m.meinke@aia.rwth-aachen.de

W. Schröder
Institute of Aerodynamics, RWTH Aachen University, Aachen, Germany

Jülich Aachen Research Alliance - High Performance Computing, RWTH Aachen University, Aachen, Germany
e-mail: office@aia.rwth-aachen.de


M.M. Resch et al. (eds.), *Sustained Simulation Performance 2017*,
DOI 10.1007/978-3-319-66896-3_9

involved, since an accurate computation of particle-laden flow requires the full resolution of the flow up to the sub-Kolmogorov scale. Only recently, access has been gained to direct particle-fluid simulations (DPFS), where all relevant scales are fully resolved without employing any models [17]. Fundamental studies have been performed and the modulation of isotropic turbulence by particles has been investigated [18, 19], which provide now a sound basis for the development of models suitable for industrial applications. A first simplification of DPFS is the direct numerical simulation (DNS), where all turbulent scales are resolved, while the particle-fluid interaction is modeled by an Euler-Lagrange model (ELM). However, DNS still requires considerable computational resources. A further simplification is provided by large-eddy simulations (LES), where large energy containing scales are resolved while models for small subgrid scales are employed mainly responsible for the dissipation.

In this contribution, a setup is developed for the comparison of DPFS, DNS, and LES, offering the possibility to use the insights gained in [18] and [19] for the development of ELM models in the framework of LES. After presenting the governing equations in Sect. 2, the numerical methods developed for the DPFS are briefly recapitulated, and an implicit LES model is introduced in Sect. 3. In Sect. 4 the latter is validated for single phase flow, and subsequently, statistics generated by LES and DNS are compared with the results of DPFS. Section 5 gives a brief conclusion emphasizing the need of further benchmark cases for a thorough analysis of the differences between the results of LES, DNS, and DPFS.

## 2 Mathematical Models

In this section, mathematical models are introduced which are capable of describing the motion of small particles suspended in a flow field. The mathematical model of the fluid phase will be given in Sect. 2.1. Thereafter, the motion of particles will be described by model equations for DPFS fulfilling the no-slip condition at particle surfaces as well as using a popular Lagrangian point particle approach in Sect. 2.2.

### 2.1 Navier-Stokes Equations

The conservation of mass, momentum, and energy in a time-dependent control volume $V$ with the surface $\partial V$ moving with the velocity $\boldsymbol{u}_{\partial V}$ may be expressed in integral form by

$$\int_{V(t)} \frac{\partial \boldsymbol{Q}}{\partial t} \, \mathrm{d}V + \int_{\partial V(t)} \bar{\boldsymbol{H}} \cdot \boldsymbol{n} \, \mathrm{d}A = \boldsymbol{0}, \tag{1}$$

where $\boldsymbol{Q} = [\, \rho_{\mathrm{f}},\ \rho_{\mathrm{f}}\boldsymbol{u}^{\mathrm{T}},\ \rho_{\mathrm{f}} E \,]^{\mathrm{T}}$ is the vector of conservative Eulerian variables and $\bar{\boldsymbol{H}}$ is the flux tensor through $\partial V$ in outward normal direction $\boldsymbol{n}$. The conservative variables are defined by the fluid density $\rho_\mathrm{f}$, the vector of velocities $\boldsymbol{u}$, and the total specific energy $E = e + |\boldsymbol{u}|^2/2$ containing the specific internal energy $e$. It is physically meaningful as well as useful for the development of numerical schemes to divide $\bar{\boldsymbol{H}}$ into an inviscid part $\bar{\boldsymbol{H}}_\mathrm{inv}$ and a viscous part $\bar{\boldsymbol{H}}_\mathrm{visc}$, where

$$\bar{\boldsymbol{H}} = \bar{\boldsymbol{H}}_\mathrm{inv} + \bar{\boldsymbol{H}}_\mathrm{visc} = \begin{pmatrix} \rho_\mathrm{f}\boldsymbol{u} \\ \rho_\mathrm{f}\boldsymbol{u}\,(\boldsymbol{u} - \boldsymbol{u}_{\partial V}) + p\bar{\boldsymbol{I}} \\ \rho_\mathrm{f}E\,(\boldsymbol{u} - \boldsymbol{u}_{\partial V}) + \boldsymbol{u}p\bar{\boldsymbol{I}} \end{pmatrix} - \frac{1}{Re}\begin{pmatrix} \boldsymbol{0} \\ \bar{\boldsymbol{\tau}} \\ \bar{\boldsymbol{\tau}}\boldsymbol{u} - \boldsymbol{q} \end{pmatrix}, \tag{2}$$

with the pressure $p$, the stress tensor $\bar{\boldsymbol{\tau}}$, the vector of heat conduction $\boldsymbol{q}$, the unit tensor $\bar{\boldsymbol{I}}$, and the Reynolds number $Re$. The latter is determined by $Re = \frac{\rho_\infty u_\infty L}{\mu_\infty}$, given the reference quantities of the density $\rho_\infty$, the velocity $u_\infty$, the length $L_\infty$, and the dynamic viscosity $\mu_\infty$. Using Stokes' hypothesis for a Newtonian fluid yields an equation for the stress tensor

$$\bar{\boldsymbol{\tau}} = 2\mu\bar{\boldsymbol{S}} - \frac{2}{3}\mu\,(\nabla\cdot\boldsymbol{u})\,\bar{\boldsymbol{I}}, \tag{3}$$

in which $\bar{\boldsymbol{S}}$ holds the rate-of-strain tensor defined as $\bar{\boldsymbol{S}} = \frac{(\nabla\boldsymbol{u} + (\nabla\boldsymbol{u})^\mathrm{T})}{2}$. The dynamic viscosity $\mu$ depends on the local thermodynamic state of the fluid. However, it can be approximately obtained by Sutherland's law

$$\mu\,(T) = \mu_\infty\left(\frac{T}{T_\infty}\right)^{3/2}\frac{T_\infty + S}{T + S}, \tag{4}$$

with S being the Sutherland temperature. Fourier's law gives the heat conduction

$$\boldsymbol{q} = -\frac{\mu}{Pr\,(\gamma - 1)}\nabla T, \tag{5}$$

using the static temperature $T$, the constant capacity ratio $\gamma = c_\mathrm{p}/c_\mathrm{v}$, the specific heat capacities $c_\mathrm{v}$ and $c_\mathrm{p}$ at constant volume and at constant pressure. The Prandtl number $Pr$ is given by $Pr = \frac{\mu_\infty c_\mathrm{p}}{k_\mathrm{t}}$ containing the thermal conductivity $k_\mathrm{t}$. The system of equations can be closed by the caloric state equation $e = c_\mathrm{v}T$ and the state equation of an ideal gas $p = \rho RT$, with $R$ being the specific gas constant.

## 2.2 *Particle Dynamics*

In this contribution, dilute suspensions of small, rigid, spherical particles with statistically negligible collisions are investigated. The volume fraction $\Phi_\mathrm{p} = V_\mathrm{p}/V$, with the volume occupied by particles $V_\mathrm{p}$ and the overall volume $V$, is small, i.e., $\Phi_\mathrm{p} \ll 1$, while the mass fraction $\psi_\mathrm{p} = M_\mathrm{p}/m_\mathrm{f}$, with the overall mass of particles $M_\mathrm{p}$ and the mass of the fluid $m_\mathrm{f}$, has a finite value, which yields an interaction between inertial particles and the smallest turbulent scales referred to as two-way coupling [4]. The linear motion of a particle $p$ with the velocity $\boldsymbol{v}_\mathrm{p}$ and mass $m_\mathrm{p}$ at the position $\boldsymbol{x}_\mathrm{p}$ is given by the relations

$$\frac{\mathrm{d}\boldsymbol{x}_\mathrm{p}}{\mathrm{d}t} = \boldsymbol{v}_\mathrm{p}, \tag{6}$$

$$m_\mathrm{p}\frac{\mathrm{d}\boldsymbol{v}_\mathrm{p}}{\mathrm{d}t} = \boldsymbol{F}. \tag{7}$$

The rotational movement $\boldsymbol{\omega}_\mathrm{p}$ of the particles may be described conveniently in a rotating frame of reference $(\tilde{x}, \tilde{y}, \tilde{z})$, which is aligned with the principal components of the particles and fixed at its center of mass, with the equation

$$\widetilde{\overline{\boldsymbol{I}}}\frac{d\widetilde{\boldsymbol{\omega}_\mathrm{p}}}{dt} + \widetilde{\boldsymbol{\omega}}_\mathrm{p} \times \left(\widetilde{\overline{\boldsymbol{I}}}\widetilde{\boldsymbol{\omega}}_\mathrm{p}\right) = \widetilde{\boldsymbol{T}}, \tag{8}$$

where $\widetilde{\overline{\boldsymbol{I}}}$ denotes the principal moments of inertia. The particle dynamics can be fully described, provided that the hydrodynamic force $\boldsymbol{F}$ and torque $\boldsymbol{T}$ acting on the particle are known. These are differently determined by DPFS and ELM, as pointed out in the following.

### 2.2.1 Direct Particle-Fluid Simulation

The full resolution of the particles establishes the no-slip condition at particle surface $\Gamma_p$, i.e., the fluid velocity on the particle surface with the particle radius $\boldsymbol{r}_\mathrm{p}$ is given by

$$\boldsymbol{u} = \boldsymbol{v}_\mathrm{p} + \boldsymbol{\omega}_\mathrm{p} \times \left(\boldsymbol{x}_\mathrm{p} - \boldsymbol{r}_\mathrm{p}\right). \tag{9}$$

Therefore, the hydrodynamic force and torque is defined by the surface integrals

$$\boldsymbol{F}_\mathrm{p} = \oint_{\Gamma_\mathrm{p}} (-p\boldsymbol{n} + \overline{\boldsymbol{\tau}} \cdot \boldsymbol{n})\, dA \tag{10}$$

$$\boldsymbol{T}_\mathrm{p} = \oint_{\Gamma_\mathrm{p}} \left(\boldsymbol{x} - \boldsymbol{r}_\mathrm{p}\right) \times (-p\boldsymbol{n} + \overline{\boldsymbol{\tau}} \cdot \boldsymbol{n})\, dA. \tag{11}$$

It should be noted that the impact of the particles on the fluid is naturally given without employing any models in contrast to ELM.

### 2.2.2 Euler-Lagrange Model

For the ELM, the no-slip condition cannot be imposed and the hydrodynamic force acting on the particles has to be modeled. Therefore, a popular simplification (e.g. [1]) of the semi-empirical Maxey-Riley equation [9], with

$$\boldsymbol{F}_{\mathrm{pp}} = 3\pi\mu d_{\mathrm{p}}(\boldsymbol{u} - \boldsymbol{v})\phi(Re_{\mathrm{p}}), \tag{12}$$

is used in this contribution, which represents the quasi-steady Stokes drag with an empirical correction function $\phi(Re_{\mathrm{p}})$ containing the particle Reynolds number $Re_{\mathrm{p}}$. However, the validity of Eq. (12) is essentially limited by the constraint $d_{\mathrm{p}}/\eta_{\mathrm{k}} \ll 1$. Specifically, with $\eta_{\mathrm{k}} \sim l_0 Re^{-3/4}$ and $l_0$ as the length scale of the largest eddy, Eq. (12) has only restricted significance for industrial and natural flow conditions which have in general a high $Re$. Additionally, the coupling force $\boldsymbol{F}_{\mathrm{pp}}$ has to be included in the momentum balance of Eq. (1) to establish the interphase coupling. Equations (1) and (12) yield a closed system of equations together with the equations of linear motion, provided that the *undisturbed* fluid velocity at the particle position $\boldsymbol{x}_{\mathrm{p}}$ may be estimated by interpolation of the *disturbed* fluid velocity at the particle position using the Eulerian velocities of the carrier flow. However, this estimate is only valid for $d_{\mathrm{p}} \ll \Delta$ [3], with $\Delta$ the grid width, which again limits the applicability of the ELM. The hydrodynamic torque is negligible for small spherical particles and may thus be safely omitted.

## 3 Numerical Methods

In this section, numerical methods for the solution of the system of equations given in Sect. 2 will be presented. First, methods for DPFS presented in [13, 15], and [17] will be briefly described. Next, an implicit LES will be introduced, which allows to control the amount of numerical dissipation added by the numerical schemes. This section will be closed with the solution schemes necessary for the ELM.

DPFS as well as the implicit LES rely on a cell-centered finite-volume formulation employing Cartesian meshes. A highly scalable efficient parallel mesh generator is available [7], where the domain decomposition is based on a weighted Hilbert curve. The inviscid fluxes $\bar{\boldsymbol{H}}_{\mathrm{inv}}$ are computed by a variant of the AUSM [8] with a modified pressure splitting proposed in [11]. Second-order accuracy is achieved via a MUSCL extrapolation routine [23], while the extrapolation uses the cell-centered gradients of the primitive variables obtained by a weighted second-order least-squares approach [17]. The viscous fluxes $\bar{\boldsymbol{H}}_{\mathrm{visc}}$ are computed by a recentering approach proposed in [2].

## 3.1 Direct Particle-Fluid Simulation

DPFS relies on an accurate computation of freely moving boundaries. This is achieved via a level-set function for the sharp representation of the boundaries and a strictly conservative numerical discretization of the cut cells at the boundaries. Using multiple level-set functions allows the resolution of particle collisions [16]. Instabilities due to arbitrary small cut cells are suppressed by an accurate interpolation scheme and conservation is ensured by a flux-redistribution technique, which also handles emerging and submerging cells due to the moving boundaries [17]. Several strategies are employed to mitigate the computational effort. First, a novel predictor-corrector Runge-Kutta scheme has been developed, which substantially reduces the overhead of remeshing and reinitialization of the solver due to the moving boundaries [17]. Next, a solution-adaptive refinement strategy generates automatically the mesh used during the solution of the flow field (cf. Fig. 1). Hence, the mesh is constantly changing since the particle positions and the flow field are different after each time step, which yields a significant load imbalance. Therefore, a dynamic load balancing method has been developed to allow the use of high-performance computers for the solution of particle-laden flows. After a predefined number of time steps, a Hilbert curve is computed on the coarsest refinement level and weighted by the number of offsprings of each cell. This yields a unique balanced domain decomposition, which can be used to redistribute the cells among the processes. Since the domain boundaries are shifted moderately, only a part of the cells in the domains have to be exchanged. Figure 2 shows a comparison of the performance for a static and dynamic domain decomposition, which has been measured for $\mathcal{O}(1000)$ particles suspended in a Taylor-Green vortex [15]. A DPFS would eventually run out of memory on a static domain decomposition, whereas a dynamic domain decomposition yields an almost constant mean wall time. The overhead for the additional communication of 6% is small compared to the speed-up gained by load-balancing.

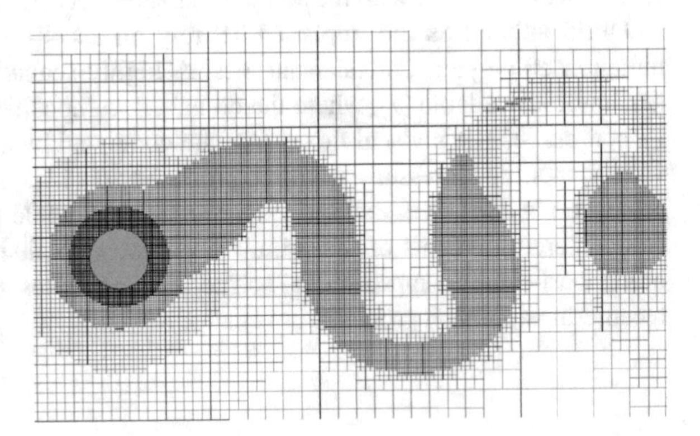

**Fig. 1** Adaptively refined grid for an elastically mounted sphere. Distances to boundaries as well as sensors for entropy gradients and vorticity control the refinement. For details on the flow case, the reader is referred to [14], while the adaptive mesh refinement is described in [6]

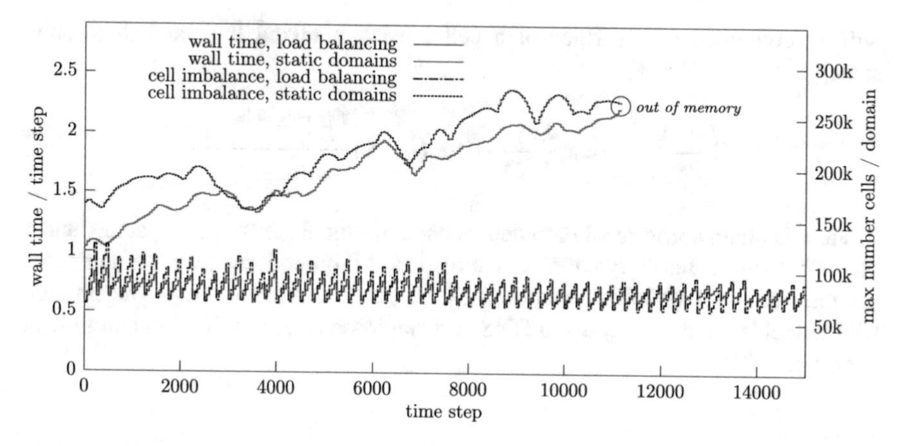


**Fig. 2** Comparison of the performance of the static and dynamic loadbalancing. Due to the preferential concentration of the particle, the load imbalance grows, and eventually exceeds the available memory, if a static domain decomposition is used. A dynamic domain decomposition yields a sawtooth-curve, where the load balancing has been performed every 250th time step

## 3.2 Implicit Large-Eddy Simulation

An AUSM scheme with a modified pressure splitting is available and it has been already shown that it is suited as an implicit LES scheme [11]. To reduce the dissipative behavior of the second-order discretization in low-Mach number flows, a modified version of the reconstruction method proposed in [22] is used. This reconstruction method alters the extrapolated velocities $u_{\mathrm{L/R}}$ at the cell faces which are needed for the AUSM by

$$u_{\mathrm{L}}^* = \frac{u_{\mathrm{L}} + u_{\mathrm{R}}}{2} + z\frac{u_{\mathrm{L}} - u_{\mathrm{R}}}{2},$$
$$u_{\mathrm{R}}^* = \frac{u_{\mathrm{L}} + u_{\mathrm{R}}}{2} + z\frac{u_{\mathrm{R}} - u_{\mathrm{L}}}{2}, \qquad (13)$$

where $u_{\mathrm{L/R}}^*$ are the altered surface velocities and $z \leq 1$ may be in general an arbitrary function. It will be chosen

$$z = min(\ 1,\ \lambda\ max(\ M_{\mathrm{r}}^{\mathrm{n}}, M_{\mathrm{l}}^{\mathrm{n}}\ )\ ), \qquad (14)$$

with the normal Mach numbers $M_{\mathrm{r,l}}^{\mathrm{n}}$ at the cell surfaces and $\lambda$ as a grid resolution dependent constant. A value of $z = 1$ recovers the original MUSCL scheme, whereas for $z$ tending to zero the surface velocities are obtained by central differencing such that velocity jumps are smoothed in low Mach number flows. The viscous fluxes are computed using a low-dissipation variation of the central scheme proposed in [2], where the normal derivatives of the normal velocity component will be computed at a surface of a cell $i$ using a mixed five- and three-point stencil

$$\left(\frac{\partial u}{\partial x}\right)_{\mathrm{i}+1/2} \approx \kappa \frac{u_{\mathrm{i}+1} - u_{\mathrm{i}}}{\delta x} + (1-\kappa)\frac{\left(\frac{\partial u}{\partial x}\right)_{\mathrm{i}} + \left(\frac{\partial u}{\partial x}\right)_{\mathrm{i}+1}}{2}, \tag{15}$$

where $\kappa$ is again a grid resolution dependent constant, $\delta x$ is the grid spacing, and $x$ only serves as an auxiliary coordinate direction. All other derivatives are computed via the five-point stencil as proposed by Berger and Aftosmis [2]. The implicit LES model is validated against a DNS of a single-phase isotropic turbulent flow in Sect. 4.1.

### 3.3 Euler-Lagrange Model

In the ELM, the particles are tracked solving Eqs. (6), (7), and (12) by a predictor-corrector scheme described in [21]. In dilute suspensions particle collisions are statistically irrelevant and thus neglected. In general, particle positions do not coincide with the cell centers and the velocity of the carrier flow "seen" by the particles has to be interpolated. Ordinary interpolation routines, however, introduce filtering errors leading to a systematic underestimate of the turbulent kinetic energy after interpolation. To avoid filtering effects, the nearest cell-centered velocity is used instead of an interpolation routine.

As described in Sect. 2, the coupling of the force in Eq. (12) has to be projected onto the grid to establish a two-way coupling. Therefore, the force is smoothly projected using the distance based weighting function

$$\boldsymbol{F}_{\mathrm{proj,i}} = \boldsymbol{F}_{\mathrm{pp}} \cdot \frac{e^{-\left(d_{\mathrm{i}}^2/(\sigma\Delta^2)\right)}}{\sum_i e^{-\left(d_{\mathrm{i}}^2/(\sigma\Delta^2)\right)}} \tag{16}$$

onto the nearest cells, with $\boldsymbol{F}_{\mathrm{proj,i}}$ the force projected on the cell $i$, $d_{\mathrm{i}}^2/\Delta^2$ the normalized distance between the cell center of the cell $i$ and the particle center, and $\sigma$ a smoothing parameter. The quantity $\sigma$ is chosen sufficiently high to avoid self-induced disturbances [10].

## 4 Results and Discussion

Isotropic particle-laden flow is examined using the numerical methods presented in Sect. 3 to solve the equations introduced in Sect. 2. The flow field of a fully periodic cube with an edge length of $L$ is initialized randomly and divergence free while fulfilling the realizability conditions [20]. To avoid compressibility effects, the Mach number was set to 0.1. The initialization procedure follows the method proposed in [12], where a prescribed energy spectrum $E(k)$ serves as initial condition with the model spectrum

$$E(k) = \left(\frac{3u_0^2}{2}\right)\left(\frac{k}{k_\mathrm{p}}^2\right)\exp\left(-\frac{k}{k_\mathrm{p}}\right), \tag{17}$$

the wave number $k = |\boldsymbol{k}|$ including the wave number vector $\boldsymbol{k}$, the peak wave number $k_\mathrm{p}$, and the initial dimensionless root-mean square velocity (rms-velocity) $u_0$. The peak wave number is chosen $k_\mathrm{p} = 4k_0$ with $k_0 = {}^{2\pi}/_{L}$. The pressure field is computed by solving the Poisson equation in spectral space as shown in [20] and the density field is obtained assuming an isothermal flow field. The initial microscale Reynolds number is set to $Re_{\lambda 0} = 79.1$. For the initialization of the LES, the energy spectrum is cut off at the highest resolvable wave number.

In the following, it will be shown that the LES is capable of predicting the single-phase isotropic turbulence correctly. Subsequently, a particle-laden case is examined and DNS as well as LES using the ELM are compared with DPFS.

## 4.1 Large-Eddy Simulation of Isotropic Turbulence

Three grid resolutions with $64^3$, $96^3$, and $128^3$ cells have been used for the LES. The findings have been compared with the results of a DNS with $256^3$ cells. Figure 3 shows the temporal development of turbulent kinetic energy using an LES with $64^3$ cells for different parameters $\lambda$ in comparison to a DNS using $256^3$, and to the original AUSM-scheme without a modification of the extrapolated velocities. In this contribution, the turbulent kinetic energy $E_k$ is normalized by its initial value $E_{k,0}$, whereas the time $t$ is normalized by the initial eddy turnover time, i.e., $t^* = t\epsilon_0/u_0^2$, with the initial viscous dissipation rate $\epsilon_0$. It can be observed that the original AUSM-scheme suffers substantially from an enhanced numerical dissipation, which can not be used as an implicit turbulence model for this flow regime. The modification offers a remedy and improves the

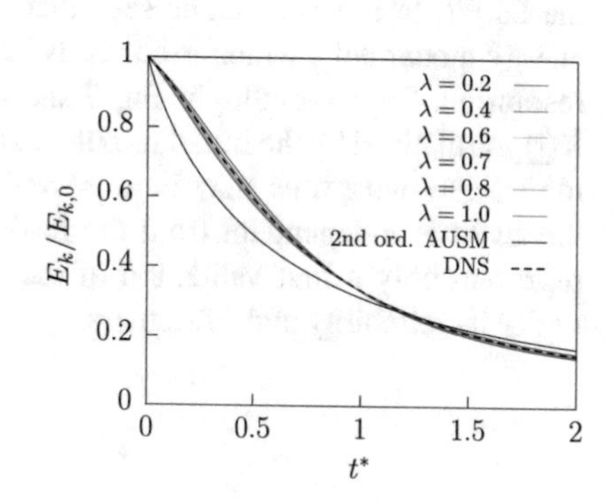


**Fig. 3** Turbulent kinetic energy using LES with the modification according to Eq. (3.2) in comparison to the original AUSM scheme and a DNS

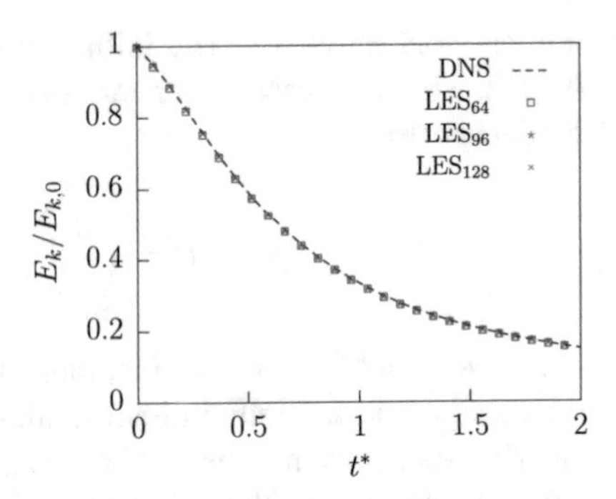


**Fig. 4** Turbulent kinetic energy using LES with the optimal parameter combination $\lambda$ and $\kappa$ for three grid resolutions

results significantly for any parameter $\lambda$. Choosing the optimal parameters for $\lambda$ and $\kappa$ yields results matching the DNS for all resolutions as can be seen in Fig. 4.

## 4.2 Turbulence Modulation by Particles

Next, the particles are induced randomly into the turbulent flow at $t_{\mathrm{i}}^* = 0.27$, which allows the turbulent flow to establish a non-linear turbulent transport (e.g. [5]). 45,000 particles with a particle density ratio $\rho_{\mathrm{p}}/\rho_{\mathrm{f}} = 1000$ and diameter on the order on the initial Kolmogorov scale, i.e., $d_{\mathrm{p}}/\eta_{\mathrm{k}} = 1.32$, are initialized with the local fluid velocity. The results of the DNS and the LES using the ELM proposed in Sect. 3.3 are validated against the benchmark results of a DPFS analyzed in [18]. An instantaneous snapshot of the flow field of a DPFS is shown in Fig. 5. Note that DPFS strongly relies on high performance computing systems, i.e., the DPFS performed in [18] required 48,000 computing cores on the Cray XC 40 of the HLRS. Moreover, the simulations using adaptive mesh refinement required about $2 \cdot 10^9$ cells, while a uniform mesh would require about $68 \cdot 10^9$ cells to resolve the flow field in the vicinity of the particles with the same accuracy.

Figures 6 and 7 show a comparison of a DNS and an LES using ELM with the DPFS. In Fig. 6 it can be seen that the particles attenuate the turbulent kinetic energy moderately, which is correctly predicted by the ELM independent from the resolution. Correspondingly, Fig. 7 shows the mean kinetic energy of the particles $K(t)$ normalized by the initial turbulent kinetic energy. A slight difference increasing moderately with time may be observed between the DPFS and the ELM, where the ELM is independent from the resolution. However, these preliminary results represent only a first validation of the ELM and further analyses are required to verify its reliability and robustness.

**Fig. 5** Instantaneous snapshot of the parallel projection of the turbulent particle-laden flow field. The structures are contours of the $\lambda_2$ criterion, whereas the *color* represents the velocity magnitude. Large vortical structures as well as particle induced structures in the vicinity of the particles are observed

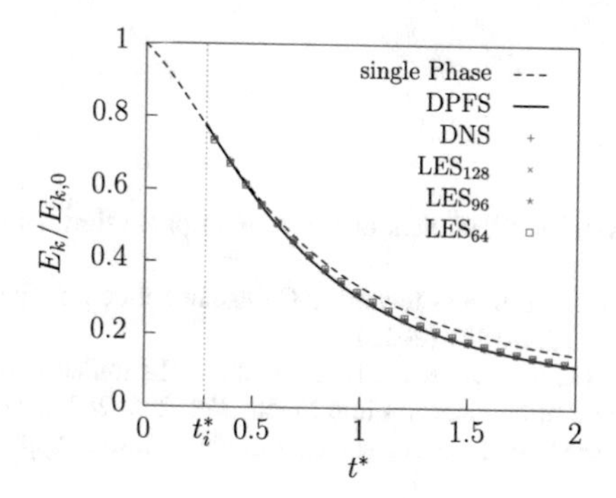


**Fig. 6** Turbulent kinetic energy of particle-laden isotropic turbulence using DPFS in comparison to DNS and LES using the ELM

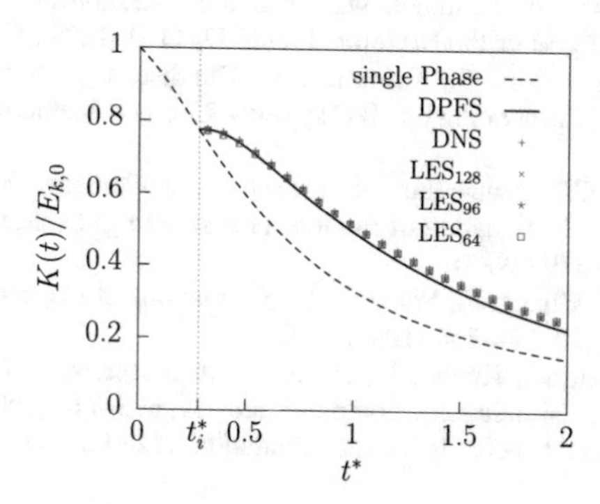


**Fig. 7** Mean kinetic energy of the particles in DPFS in comparison to LES and DNS using the ELM

## 5 Conclusion

A setup has been presented for the validation of LES and DNS using an ELM model against benchmark results generated via DPFS. Therefore, an implicit LES formulation is employed which facilitates the use of ELM for different grid resolutions. Preliminary results show only slight deviations between the DPFS and the ELM for all grid resolutions. However, this behavior is certainly dependent on various parameters. Therefore, the validation of the ELM requires more benchmark cases using DPFS. In particular, it is planned to study the turbulence modulation by larger spherical particles, non-spherical particles, and non-isothermal particles.

**Acknowledgements** This work has been funded by the German Research Foundation (DFG) within the framework of the SFB/Transregio 129 "Oxyflame" (subproject B2). The support is gratefully acknowledged. Computing resources were provided by the High Performance Computing Center Stuttgart and by the Jülich Supercomputing Center (JSC) within a Large-Scale Project of the Gauss Center for Supercomputing (GCS).

## References

1. Balachandar, S., Eaton, J.K.: Turbulent dispersed multiphase flow. Annu. Rev. Fluid Mech. **42**, 111–133 (2010)
2. Berger, M., Aftosmis, M.: Progress towards a Cartesian cut-cell method for viscous compressible flow. AIAA Paper 2012-1301 (2012)
3. Boivin, M., Simonin, O., Squires, K.D.: Direct numerical simulation of turbulence modulation by particles in isotropic turbulence. J. Fluid Mech. **375**, 235–263 (1998)
4. Elghobashi, S.: On predicting particle-laden turbulent flows. Appl. Sci. Res. **52**, 309–329 (1994)
5. Elghobashi, S., Truesdell, G.: On the two-way interaction between homogeneous turbulence and dispersed solid particles. I: turbulence modification. Phys. Fluids A **5**, 1790–1801 (1993)
6. Hartmann, D., Meinke, M., Schröder, W.: An adaptive multilevel multigrid formulation for Cartesian hierarchical grid methods. Comp. Fluids **37**, 1103–1125 (2008)
7. Lintermann, A., Schlimpert, S., Grimmen, J., Günther, C., Meinke, M., Schröder, W.: Massively parallel grid generation on HPC systems. Comput. Methods Appl. Mech. Eng. **277**, 131–153 (2014)
8. Liou, M.-S., Steffen, C.J.: A new flux splitting scheme. J. Comput. Phys. **107**, 23–39 (1993)
9. Maxey, M.R., Riley, J.J.: Equation of motion for a small rigid sphere in a nonuniform flow. Phys. Fluids **26**, 883–889 (1983)
10. Maxey, M., Patel, B., Chang, E., Wang, L.-P.: Simulations of dispersed turbulent multiphase flow. Fluid Dyn. Res. **20**, 143–156 (1997)
11. Meinke, M., Schröder, W., Krause, E., Rister, T.: A comparison of second-and sixth-order methods for large-eddy simulations. Comput. Fluids **31**, 695–718 (2002)
12. Orszag, S.A.: Numerical methods for the simulation of turbulence. Phys. Fluids **12**, II-250 (1969)
13. Schneiders, L., Hartmann, D., Meinke, M., Schröder, W.: An accurate moving boundary formulation in cut-cell methods. J. Comput. Phys. **235**, 786–809 (2013)
14. Schneiders, L., Meinke, M., Schröder, W.: A robust cut-cell method for fluidstructure interaction on adaptive meshes. AIAA Paper 2013-2716 (2013)

15. Schneiders, L., Grimmen, J.H., Meinke, M., Schröder, W.: An efficient numerical method for fully-resolved particle simulations on high-performance computers. PAMM **15**, 495–496 (2015)
16. Schneiders, L., Günther, C., Grimmen, J.H., Meinke, M.H., Schröder, W.: Sharp resolution of complex moving geometries using a multi-cut-cell viscous flow solver. AIAA 2015-3427 (2015)
17. Schneiders, L., Günther, C., Meinke, M., Schröder, W.: An efficient conservative cut-cell method for rigid bodies interacting with viscous compressible flows. J. Comput. Phys. **311**, 62–86 (2016)
18. Schneiders, L., Meinke, M., Schröder, W.: Direct particle-fluid simulation of Kolmogorov-length-scale size particles in decaying isotropic turbulence. J. Fluid Mech. **819**, 188–227 (2017)
19. Schneiders, L., Meinke, M., Schröder, W.: On the accuracy of Lagrangian point-mass models for heavy non-spherical particles in isotropic turbulence. Fuel **201**, 2–14 (2017)
20. Schumann, U., Patterson, G.: Numerical study of pressure and velocity fluctuations in nearly isotropic turbulence. J. Fluid Mech. **88**, 685–709 (1978)
21. Siewert, C., Kunnen, R., Schröder, W.: Collision rates of small ellipsoids settling in turbulence. J. Fluid Mech. **758**, 686–701 (2014)
22. Thornber, B., Mosedale, A., Drikakis, D., Youngs, D., Williams, R.J.: An improved reconstruction method for compressible flows with low Mach number features. J. Comput. Phys. **227**, 4873–4894 (2008)
23. Van Leer, B.: Towards the ultimate conservative difference scheme. V. A second-order sequel to Godunov's method. J. Comput. Phys. **32**, 101–136 (1979)

# Coupled Simulation with Two Coupling Approaches on Parallel Systems

**Neda Ebrahimi Pour, Verena Krupp, Harald Klimach, and Sabine Roller**

**Abstract** The reduction of noise is one of the challenging tasks in the field of engineering. The interaction between flow, structure, and an acoustic field involves multiple scales. Simulating the whole domain with one solver is not feasible and out of range on todays supercomputer. Since the involving physics appear on different scales, the effects can be spatially separated into different domains. The interaction between the domains is realised with coupling approaches via boundaries. Different interpolation methods at the coupling interfaces are reviewed in this paper. The methods include the Nearest-Neighbor Interpolation (first order), the Radial-Basis Function (second order) as well as the direct evaluation of the state representation at the requested points (arbitrary order). We show which interpolation method provides less error, when compared to the monolithic solution of the result. We present how the two coupling approaches *preCICE* and *APESmate* can be used. The coupling tool *preCICE* is based on a black box coupling, where just the point values at the surface of the coupling domains are known. In contrast *APESmate* has knowledge about the numerical schemes within the domain. Thus, *preCICE* needs to interpolate values, while *APESmate* can evaluate the high order polynomials of the underlying Discontinous Galerkin scheme. Hence, the *preCICE* approach is more generally applicable, while the *APESmate* approach is more efficient, especially in the context of high order schemes.

## 1 Introduction

With increasing computational resources also the idea of simulating more complex and larger simulations gets more and more important, since they allow a better understanding of physical phenomena and the optimisation in product design. In the recent years the energy turnaround in renewable resources gained more and more popularity, which lead to an increasing number of e.g. wind turbines. Wind turbines emit noise, which is caused by the rotor-wind interaction, where turbulent

---

N. Ebrahimi Pour (✉) • V. Krupp • H. Klimach • S. Roller
University of Siegen, Adolf-Reichwein-Str. 2, Siegen, Germany
e-mail: neda.epour@uni-siegen.de; verena.krupp@uni-siegen.de; harald.klimach@uni-siegen.de; sabine.roller@uni-siegen.de


M.M. Resch et al. (eds.), *Sustained Simulation Performance 2017*,
DOI 10.1007/978-3-319-66896-3_10

flow appears. With the increasing number and size of the turbines, the reduction of noise becomes important, due to the noise emission in the range of hundreds of meters or up to a few kilometres. The interaction of flow, structure and acoustic field (FSA) has to be studied in more detail, to improve the understanding of these factors, that can help to reduce the noise propagation. Solving the whole domain with a single equation (monolithic) up to the smallest scales is not feasible and still out of reach on todays supercomputers. Therefore we use partitioned coupling, where the physical space is divided into smaller domains and each of the subdomains covers a dedicated physical setup, as the different physical effects appear on different scales. Hence the separation of the domains allows different numerical treatments of the large and complex problem, which is than feasible as different equations as well as mesh resolutions can be used for each of them.

For the communication and the data exchange between the subdomains coupling approaches are used. They are based on a heterogeneous domain decomposition, where the different domains are connected to each other via boundary conditions. Considering compressible flows, we make use of explicit time stepping schemes that enable a straightforward data exchange at the coupling interfaces. Applying a proper data exchange at the coupling boundaries allows the deployment of various schemes and equations in each domain. Thus the numerical approximation can be adapted to the requirements of each domain. Especially for the acoustic far field, where no or few obstacles are present and only the wave propagation has to be considered, high order methods with low dissipation and dispersion are beneficial. Therefore the high order modal Discontinuous Galerkin Method (DGM) [5] is considered.

In this paper we investigate the two different coupling approaches. One coupling approach is a black-box tool, which allows the coupling of different solvers without any knowledge of each discretisation, while the other one is an integrated approach, which has knowledge about the underlying scheme and makes use of that. At first two interpolation methods: Nearest-Neighbor (NN) and Radial-Basis Functions (RBF) with and without providing equidistant points, provided by the coupling tool *preCICE* [1, 2] are presented. Afterwards the direct evaluation of the state representation at requested points, provided by the coupling approach *APESmate* [6] is introduced.

The final section is devoted to simulation results with a small academic testcase, in order to compare the different interpolation methods and the resulting error. Finally we come to an end, by concluding our results.

## 2 Data Mapping Methods

This section describes the different interpolation methods as well as the evaluation of the state variables, in order to exchange point values at the coupling interfaces. Therefore first the interpolation methods in the coupling approach *preCICE* are introduced, afterwards the integrated coupling approach *APESmate* is presented.

## 2.1 Interpolation

*preCICE* is an multi-solver, which allows the coupling of different solvers, considering them as a *black-box*. Thus it has no information about the discretisation of each solver, while exchanging input and output values via the coupling interfaces using coupling points. For the exchange of values at requested points, the coupling tool provides different interpolation methods, in this paper, we are going to review three of them. More information about *preCICE* can be found in [1] and [9]. Since the domains, which are involved in the coupling, request point values located at arbitrary positions at the coupling interface, hence providing values has to be done on those requested point positions. Therefore interpolation methods are necessary to compute values from one domain to the other.

The easiest applicable interpolation method is the *Nearest-Neighbor* (NN) interpolation (see Fig. 1a) [3]. In order to use this method, the solver does not have to provide any information beside the variable values at its exchange points. Coupling from domain B to A, this method searches for the closest point on B and copies the value to the requested point on A. If more than one point of domain A is in the near of one point in domain B, than all those points get the same value. Therefore this interpolation method is just first order accurate and useful, when having a matching coupling.

The *Nearest-Projection* (NP) method (see Fig. 1b) looks for the closest neighboring point of domain A among domain B, while computing the projection point of the point in B on the point in domain A. Thus a linear equation has to be solved, which leads to a second order accuracy. In order to make use of this method, the solver has to provide neighborhood information, in form of triangles or edges [3].

A second order accurate method, where no neighborhood information is necessary, is the *Radial-Basis-Function* method (RBF).

$$g(x) = \sum_{i=1}^{N_B} \gamma_i \cdot \phi(||x - x_i||) + q(x) \qquad (1)$$

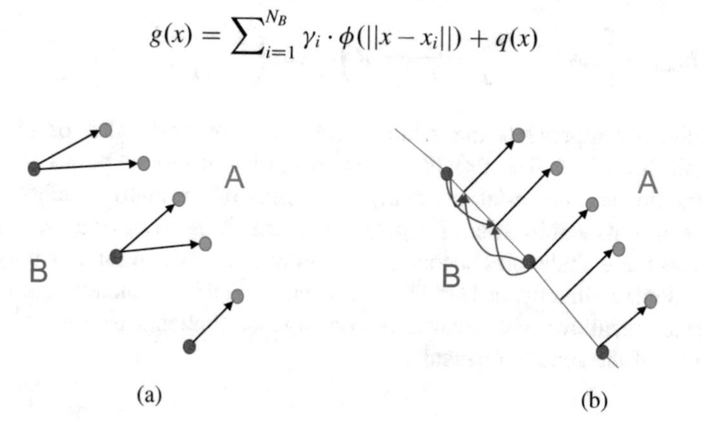


**Fig. 1** (**a**) Nearest-Neighbor, (**b**) Nearest-Projection interpolation [3]

For the mapping from domain B to A, the radial-basis-function creates a global interpolant on B, which is evaluated on A. The basis are radially symmetric basis functions, which are centred at the coupling points of domain B. Equation (1) presents the equation, which has to be solved internally by *preCICE* for the RBF. To make sure, that constant and linear functions are interpolated exactly, the basis has an additional first order global polynomial $q(x)$. The variable $\phi$ is the basis function, chosen by the user. In this paper we use the Gaussian function as basis function for the interpolation [3]. Using this basis function requires to predefine a shape-parameter $s$ (see Eq. (2)), which defines the width of the Gaussian function.

$$s = \frac{\sqrt{-\ln(10^{-9})}}{m \cdot h} \tag{2}$$

Where $m$ is the number of points which has to be covered by the Gaussian function and $h$ the average distance between the points. It has to be mentioned, that the distribution of the coupling points has a major influence on the quality of the simulation results as well as on the convergence of the linear equation system, which has to be solved. Different situations can be thought of, where the sampling points on a domain interface are non-equidistant. One is the use of high order Discontinous Galerkin (DG) schemes, where the sampling points are the Gaussian integration points which are not equally distributed. While at the corners of the cells the points are more concentrated, in the middle of the cell the distance is much larger. The size of the shape-parameter has an great impact on the convergence of the system, while selecting a high $m$ leads to an increase in the quality of the simulation, it also influences the condition number, which also increases [7]. In the DG context, the variable $h$ is set to the maximum distance between the points, to also cover the points in the middle of the cell, which have a larger distance to each other. The distance is calculated by considering the distance between the Chebysheve nodes.

$$h_{max} = \left[\cos\left(\frac{2 \cdot (n_h + 1) - 1}{2 \cdot nO} \cdot \pi\right) - \cos\left(\frac{2 \cdot n_h - 1}{2 \cdot nO} \cdot \pi\right)\right] \cdot \frac{dx}{2} \tag{3}$$

The value $nO$ represents the scheme order, $n_h$ the half value of the scheme order and $dx$ the cell size. Providing non-equidistant sampling points for the RBF reconstruction leads to instability of the convergence of the matrix, while equidistant points lead to its stabilisation. Keeping that in mind, the need of a method, which overcomes these challenges becomes more important. Thus evaluating polynomials for each domain directly, instead of using an additional interpolation method for the reconstruction, allows to overcome the convergence challenge as well as increasing the quality of the simulation results.

## 2.2 *Data Mapping by Evaluation*

The integrated coupling approach *APESmate* [6] is implemented in our *APES* framework [8, 10], thus it has access to solver specific data. Therefore exchanging data at arbitrary number of exchange points with our DGM solver *Ateles* at the coupling interface can be realised by the direct evaluation of the polynomial representations at the requested points. Hence no additional interpolation is necessary. One of the main beneficial of our coupling approach is, that with a higher order a higher accuracy in the context of simulation errors can be obtained [6].

# 3 Results

This section deals with the simulation results, when using the two different coupling methods, interpolation and evaluation. Therefore we use a Gaussian density pulse, which travels from the left domain to the right domain, to compare the different methods. We created three different testcases, (a) matching, (b) non-matching with same number of coupling points and (c) non-matching with different number of coupling points. For the simulation we change the cell size and the scheme order. The matching testcase (a) has on both domains the same number of cells as well as the same scheme order, thus the number of coupling points are also the same. For non-matching testcase (b) the left domain is kept same as for (a), while on the right domain a two times greater cell size and scheme order has been chosen, which still results in the same number of coupling points as for the left domain. Testcase (c) is also a non-matching testcase, here the left domain is again the same as in (a) and (b), while the right domain has a four times bigger cell size and the scheme order is equal to the right domain in testcase (b).

## 3.1 *Configuration of the Simulation*

For our testcases we provide a $4 \times 4$ plane, which is divided into two domains. We solve both domains with the nonlinear Euler equation and choose a Gaussian density pulse, which travels from the left domain to the right, due to the advection of the flow in x-direction. The amplitude of the pulse is set to 1.0 and a halfwidth of 0.316. The pressure $p$ is set to 8.0 and the density $\rho$ to 1.0. The velocity has a constant value of $\mathbf{v} = [12.5, 0.0, 0.0]$. Figure 2 presents the point distribution in the cells for all investigated testcases, when using DG. Table 1 provides a short overview of the investigated testcases with the different configurations.

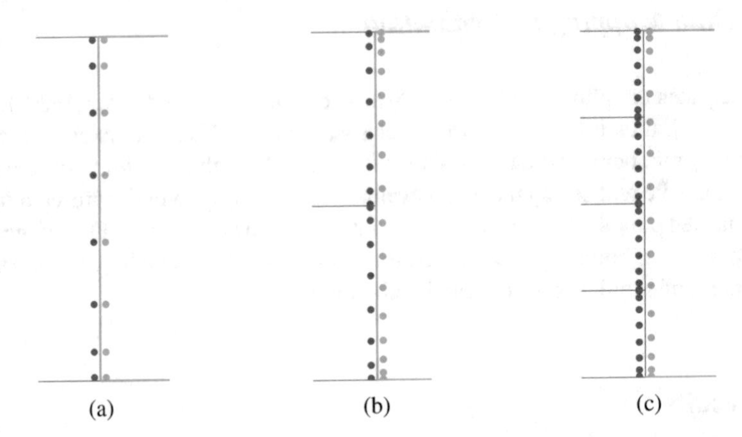


**Fig. 2** Point distribution in cells, when using DGM: (**a**) matching testcase a, (**b**) non-matching testcase b and (**c**) non-matching testcase c

**Table 1** Three testcases for the investigation of the interpolation methods

| | Testcase a matching | | Testcase b non-matching | | Testcase c non-matching | |
|---|---|---|---|---|---|---|
| | Left | Right | Left | Right | Left | Right |
| Number of cells | 512 | 512 | 512 | 256 | 512 | 128 |
| Number of coupling points | 128 | 128 | 128 | 128 | 128 | 64 |
| Scheme order | 8 | 8 | 8 | 16 | 8 | 16 |

## 3.2 *Coupled Simulation Results*

As the investigated testcase is small enough, we can obtain a monolithic simulation, running the entire testcase in one single non-split domain, and use this as the reference solution without coupling error (Fig. 3). Then we split the domain into halves, while keeping the same cell size and order. I.e. all differences between splitted and monolithic simulation results are due to the coupling error. In the following, we then change the settings in the right domain, such as to adopt the scheme order and cell size to the needs of the domain. Thus, additional errors are introduced now, due to the non-matching conditions. For the monolithic simulation, where we consider the settings as for testcase a, the error is computed from the difference between the result of the simulation and the analytical solution [6]. The error in the middle area of the domain is due to the shape of the pulse and its location at the beginning of the simulation. Since the pulse was located in the striking distance of the left boundary, oscillations appear, which traveled with the pulse through the entire domain. Which does not have any influence for the comparison of the different methods. Figure 4 shows the pulse after passing the interface for all three testcases using the RBF interpolation. As can be recognise no significant

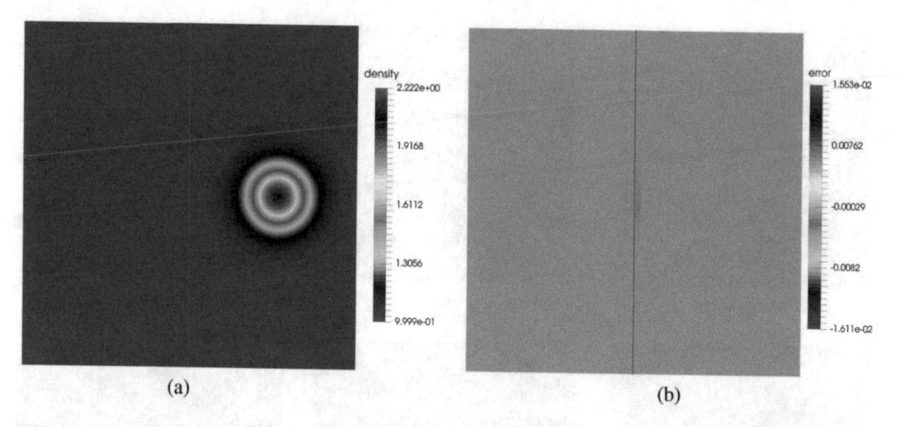


**Fig. 3** (**a**) Monolithic solution of the simulation and (**b**) Error of the simulation

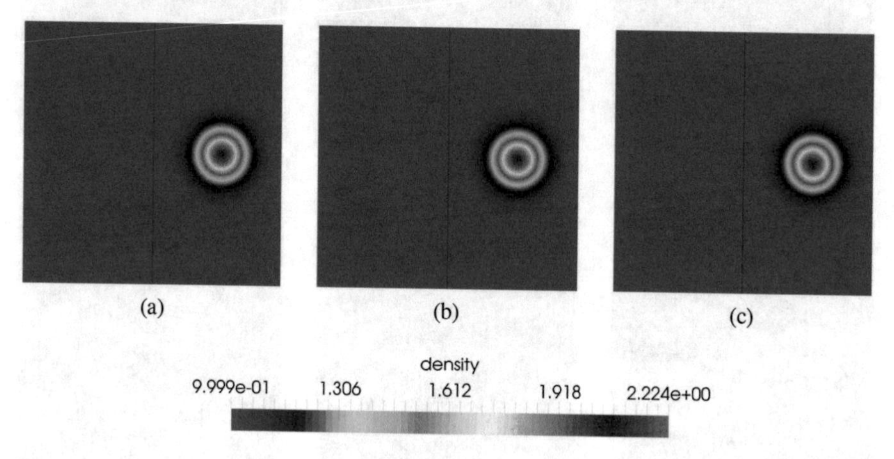


**Fig. 4** Solution of the Gaussian density pulse, which has traveled from the left domain to the right: (**a**) Testcase a, (**b**) Testcase b and (**c**) Testcase c

change of the pulse can be observed. To clarify which method shows the lowest error, we investigate the simulations in more detail. The error for the following coupled simulations are calculated from the difference between the coupled solution and the monolithic solution Fig. 3a. Thus this should just provide the error, which is due to the coupling of the domains, using interpolation or the evaluation of the polynomials respectively. In Fig. 5, a stronger impact of the matching and non-matching setup is visible. For the NN interpolation, it becomes apparent, that the error has an increasing behaviour, when having an increasing non-matching coupling interface (see Fig. 5b, c). For the matching coupling interface Fig. 5a, the error is similar to the monolithic solution, which is due to the fact, that the points on both sides coincide. Thus, the NN interpolation is a pure injection, no interpolation error is introduced. For the RBF interpolation we have to compute the shape-parameter (Sect. 2) for each domain before running the simulation. Thus

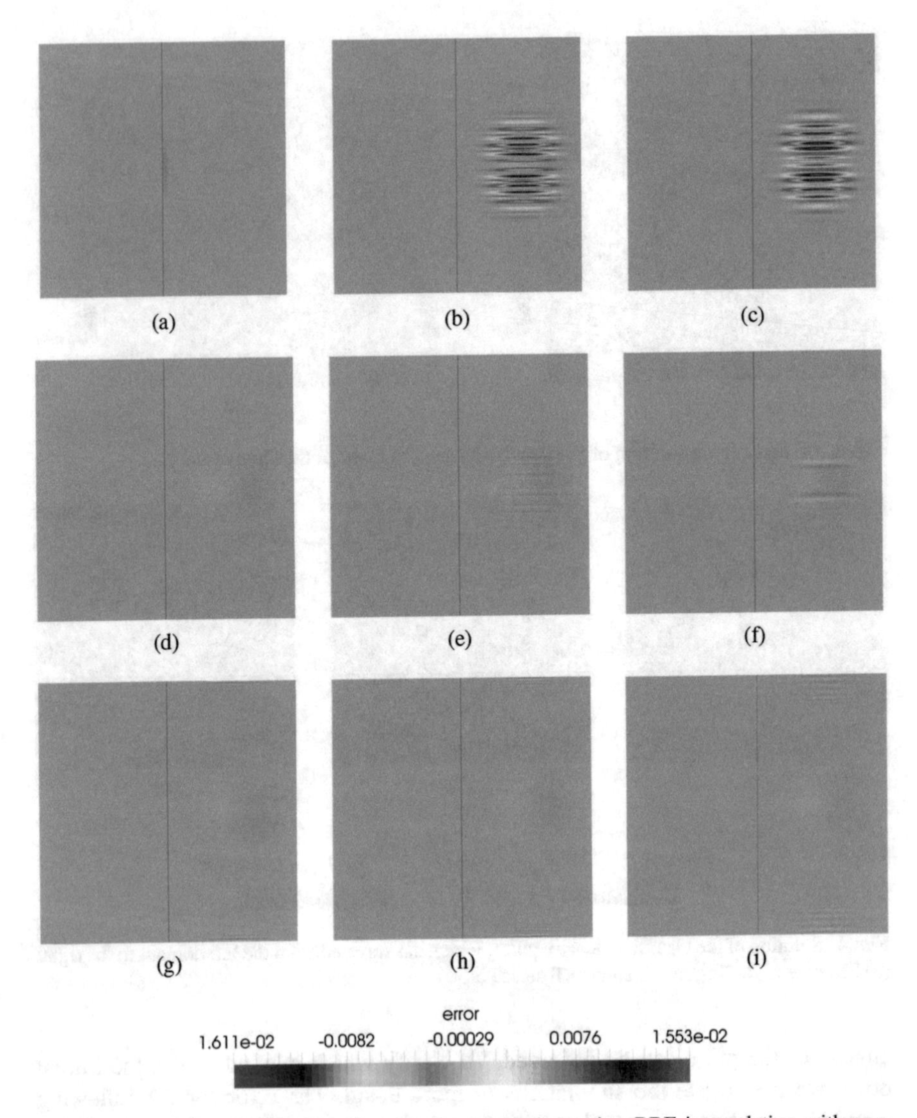


**Fig. 5** Error of the traveled Gaussian density pulse when using RBF interpolation with non-equidistant point distribution: (**a**) Testcase a with NN, (**b**) Testcase b with NN, (**c**) Testcase c with NN, (**d**) Testcase a with RBF non-equidistant Points, (**e**) Testcase b with RBF non-equidistant Points, (**f**) Testcase c with RBF non-equidistant Points, (**g**) Testcase a with RBF equidistant Points, (**h**) Testcase b with RBF equidistant Points and (**i**) Testcase c with RBF equidistant Points

for each testcase the maximal distance between the non-equidistant points has to be determined. Considering the results in Fig. 5d–f, the error, when using the RBF interpolation also increases with stronger non-matching coupling interfaces (Fig. 5e, f), while for the matching testcase Fig. 5d the error is similar to the monolithic solution. Taking Table 2 into account, we can recognise, that stronger non-matching

**Table 2** Computed shape-parameter for non-equidistant point distribution

| Testcase | $h_{max}$ Left | $h_{max}$ Right | $m$ Left | $m$ Right | $s$ Left | $s$ Right |
|---|---|---|---|---|---|---|
| a | 0.0244 | 0.0244 | 4 | 4 | 46.642 | 46.642 |
| b | 0.0244 | 0.0245 | 4 | 3 | 46.642 | 61.936 |
| c | 0.0244 | 0.0245 | 4 | 2 | 46.642 | 46.452 |

**Table 3** Computed shape-parameter for equidistant point distribution

| Testcase | $h_{max}$ Left | $h_{max}$ Right | $m$ Left | $m$ Right | $s$ Left | $s$ Right |
|---|---|---|---|---|---|---|
| a | 0.0156 | 0.0156 | 7 | 7 | 41.621 | 41.621 |
| b | 0.0156 | 0.0156 | 7 | 7 | 41.621 | 41.621 |
| c | 0.0156 | 0.0313 | 7 | 7 | 41.621 | 20.810 |

interfaces lead to a decreasing number of points $m$, which can be covered by the Gaussian function. Additional in [7] a more detailed study pointed out, that non-equidistant point distribution leads to instability of the system, thus to the not convergence of the matrix. Therefore it was suggested to consider equidistant points, to stabilise the system, hence aiming for a faster convergence. Thus in our next simulations, we provide equidistant points to *preCICE*, while asking for point values, which are non-equidistant distributed. Since we make use of the modal DG, providing and asking *preCICE* for equidistant point values would lead to higher computational effort, due to additional transformation from points to polynomials. Again we calculated according to Eq. (2) the shape-parameter for the new setup. As in Table 3 pointed, the variable $m$, can be chosen much higher. Furthermore the system converged much faster, which leads to the decreasing of computational effort. The simulation results for the equidistant point distribution illustrate for all testcases oscillations near the upper and lower boundaries, which increase with stronger non-matching coupling interfaces (see Fig. 5h, i). The oscillations did not appear, using the NN and the RBF method with non-equidistant points. This behaviour is due to the *Runge's* phenomenon [4], which appears, when using high order polynomials over equidistant interpolation points. Thus providing equidistant points for the interpolation leads to the stabilisation of the system, but decreases the quality of the simulation results. In Fig. 6 as well as in Fig. 7 the solution for all testcases show the same behaviour, thus even for the strongest non-matching testcase Fig. 7c, the error is comparable to the matching testcase Fig. 7a. Besides the visualisation of the simulation results, we also consider the L2error for the simulations to have a better comparison between the different methods. The presented Table 4 gives an overview, which method allows the lowest error, when running coupled simulations. Therefore we just consider the right domain, where the pulse reaches its final position, after passing the coupling interface. From Table 4 it can be pointed out, that for the RFB with non-equidistant points as well as for the NN interpolation with stronger non-matching coupling interfaces (c) also the error increases, while for NN this increase is distinguished. The RBF with equidistant points show already a high error for testcase (a), while a significant trend as for RBF

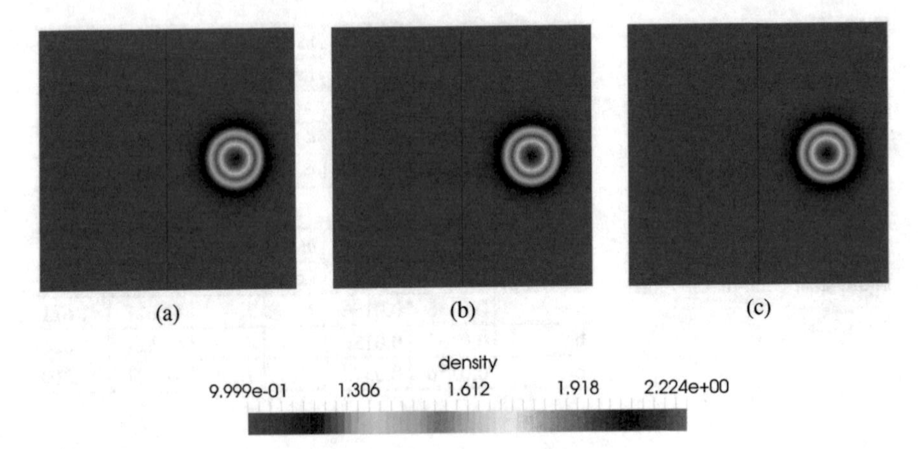


**Fig. 6** Solution of the Gaussian density pulse, which has traveled from the left domain to the right using *APESmate*: (**a**) Testcase a, (**b**) Testcase b and (**c**) Testcase c

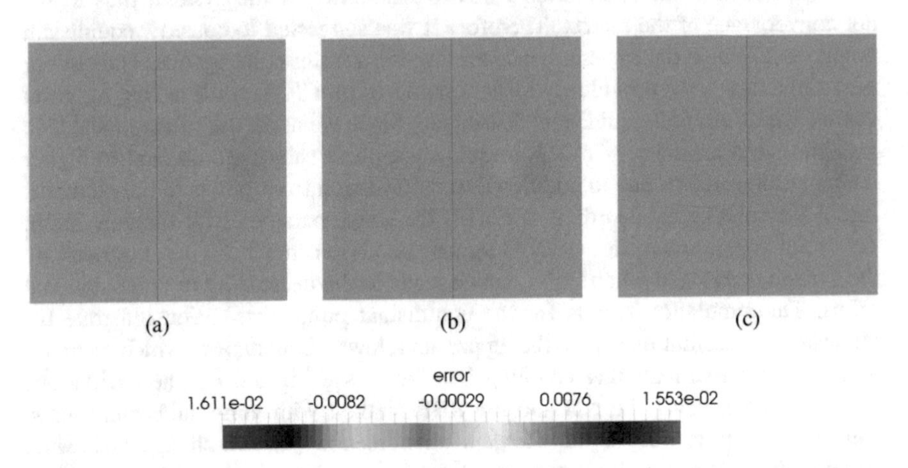


**Fig. 7** Error of the traveled Gaussian density pulse when using *APESmate*: (**a**) Testcase a, (**b**) Testcase b and (**c**) Testcase c

**Table 4** Comparison of the L2error for the different methods

| $\times 10^{-3}$ | a | b | c |
|---|---|---|---|
| Nearest-Neighbor | 1.606 | 27.774 | 52.344 |
| Radial-Basis-Function: non-equidistant points | 1.606 | 1.986 | 2.326 |
| Radial-Basis-Function: equidistant points | 1.606 | 5.101 | 3.485 |
| APESmate | 1.606 | 1.593 | 1.599 |
| Monolithic | 1.05870 | | |

with non-equidistant points and for the NN can not be observed. The increase of error values in testcase (b) and the decrease for testcase (c) are due to the changes of order and cell size in testcase (b), where the number of non-equidistant points in the corners of a cell is more dominant for a 16th order than for a 8th order simulation. The error decreases for the testcase (c), since the cell size on the right domain is two times larger than in testcase b), thus less cells are involved in the coupling, which also decreases the error. The results of *APESmate* are noticeable, even for testcase (c), where all interpolation methods provided by *preCICE* show a high error, the simulation results of our coupling approach shows the smallest and the most stabile error over all simulations.

Beside the error of the simulation results, the performance of each method is an important factor, which allows the fast computation of the simulation. Therefore also investigate the performance of the used methods.

## 3.3 Performance of the Mapping Methods

For the performance runs we consider the same settings as in Sect. 3.1, while using testcase c as a three dimensional setup. The left domain has 16,384 cells and the right domain 256. The number of coupling points at the coupling interface is 65,536 and 16,384 respectively. Since the left domain has just 256 cells, the number of cores is chosen according to this limitation. Figure 8 presents the time for the simulation

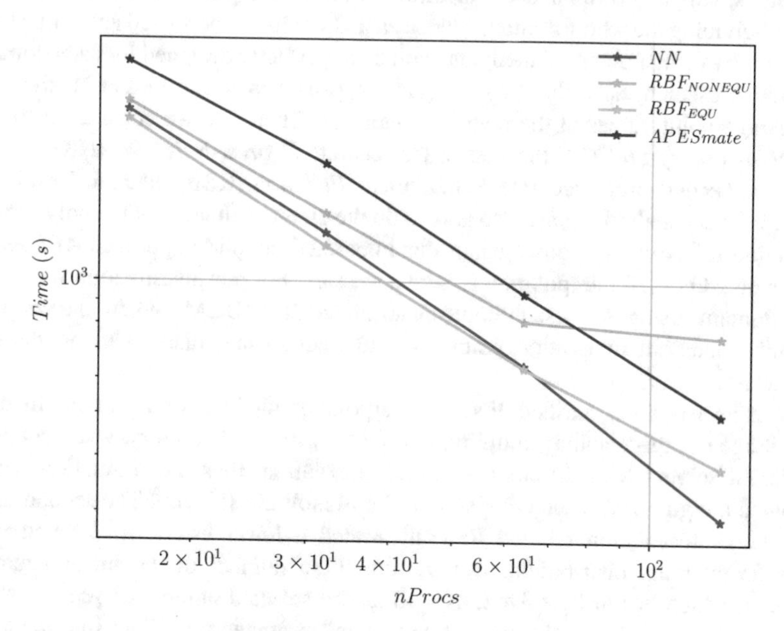


**Fig. 8** Strong scaling of the different methods

with the different methods, while changing the number of cores from 16 to 128. It is clearly visible, that the interpolation method NN, is the fastest in the computation, which is as expected, since this method just copies data from one domain to the other. But we also have to keep in mind, that apart from the fast computation, this method provides the highest error, when having non-matching coupling interfaces (see Fig. 5c). A closer look at the RBF method with different point distribution illustrates also here, that the simulation with equidistant point distribution ($RBF_{EQU}$) behaves as expected faster than the non-equidistant point distribution, since the equidistant points reduce the condition number and thus the computational time. In contrast the performance of the non-equidistant point distribution ($RBF_{NONEQU}$) gets flat, with increasing number of cores. Since our integrated approach *APESmate* evaluates polynomials for the coupling, the computation is higher, when compared to the interpolation methods provided by *preCICE*. But as was shown before the simulation results (see Fig. 7) is outstanding and in the same magnitude range as for the monolithic run.

## 4 Conclusion

The simulation of multi-scale problems is still a challenging task in the engineering field. Solving these problems with one approach is still expensive, thus a more feasible strategy is required. The partitioned coupling is one of the most promising methods, which allows the decomposition of the whole problem into smaller ones, by subdividing the whole domain. Hence each domain can be solved according to its physics by using numerical methods, which are perfectly designed for each domain.

We presented how the two coupling approaches use different methods to exchange point values at the coupling interface. Therefore we considered for the external library *preCICE* the first order accurate *Nearest-Neighbor* (NN) method and the second order accurate *Radial*-Basis-*Function* (RBF) method. Beside the interpolation method, we are also able to do the direct evaluation of the polynomials at requested coupling points, using our integrated coupling approach *APESmate*. Thus no additional interpolation is here necessary. For our investigation we solved our domains using the Discontinues-Galerkin Method (DGM), where the coupling points (Gaussian integration points) are not equidistant distributed on the cell surface.

Our investigation clarified, that the interpolation method NN is not qualified for the usage of non-matching coupling interfaces, which is necessary, when coupling different solvers with different numerical resolution, thus the simulation results showed a high error, when compared to the monolithic solution. The second order accurate interpolation method RBF illustrated a lower error, when using non-equidistant point distribution. But the condition number of the linear equation system, which has to be solved, as well as the selected number of points, which have to be covered by the basis function is unsatisfactory. Thus the condition number increases with higher scheme order and at the same time the computational effort. In order to decrease the condition number and stabilise the system, we provided *preCICE* equidistant points at the coupling interface, while asking for non-equidistant distributed points. By applying the equidistant points for the interpolation, we were able to increase the number of points, which has to be covered by the Gaussian function (basis function). Furthermore the condition number decreases and thus the stabilisation of the system could be archived. But taking also the quality of the simulation results into account, we could recognise, that oscillation occur at the lower and upper boundary, which we could not observe before, when using the NN or the RBF method with non-equidistant point distribution. These oscillations appear, due to the *Runge's* phenomenon, when using equidistant points for the interpolation of non-equidistant point distribution. The results of the simulations, when using our integrated approach *APESmate*, depict the lowest error, when compared to the monolithic solution. Even for the non-matching coupling interface, where the interpolation methods in *preCICE* show the highest error, our approach illustrated an outstanding behaviour, by having an almost constant L2error for all simulations, which is in the same magnitude range as the monolithic one.

**Acknowledgements** The financial support of the priority program 1648 - Software for Exascale Computing 214 (www.sppexa.de) of the German Research Foundation. The performance measurements were performed on the Supermuc supercomputer at Leibniz Rechenzentrum (LRZ) der Bayerischen Akademie der Wissenschaften. The authors wish to thank for the computing time and the technical support.

## References

1. Bungartz, H.J., Lindner, F., Gatzhammer, B., Mehl, M., Scheufele, K., Shukaev, A., Uekermann, B.: preCICE - a fully parallel library for multi-physics surface coupling. Comput. Fluids **141**, 250–258 (2015)
2. Bungartz, H.J., Lindner, F., Gatzhammer, B., Mehl, M., Scheufele, K., Shukaev, A., Uekermann, B.: preCICE - a fully parallel library for multi-physics surface coupling. Comput. Fluids 1 (2016). http://dx.doi.org/10.1016/j.compfluid.2016.04.003
3. Bungartz, H.J., Lindner, F., Mehl, M., Scheufele, K., Shukaev, A., Uekermann, B.: Partitioned fluid-structure-acoustics interaction on distributed data – coupling via preCICE. In: H.J. Bungartz, P. Neumann, E.W. Nagel (eds.) Software for Exa-scale Computing – SPPEXA 2013–2015. Springer, Berlin, Heidelberg (2016)
4. Fornberg, B., Zuev, J.: The Runge phenomenon and spatially variable shape parameters in RBF interpolation. Comput. Math. Appl. **54**(3), 379–398 (2007). http://dx.doi.org/10.1016/j.camwa.2007.01.028
5. Hesthaven, J.S., Warburton, T.: Nodal Discontinuous Galerkin Methods: Algorithms, Analysis, and Applications, 1st edn. Springer Publishing Company, Incorporated, New York (2007)
6. Krupp, V., Masilamani, K., Klimach, H., Roller, S.: Efficient coupling of fluid and acoustic interaction on massive parallel systems. In: Sustained Simulation Performance 2016, pp. 61–81 (2016). doi:10.1007/978-3-319-46735-1_6
7. Lindner, F., Mehl, M., Uekermann, B.: Radial basis function interpolation for black-box multiphysics simulations. In: Papadrakakis, M., Schrefler, B., Onate, E. (eds.) VII International Conference on Computational Methods for Coupled Problems in Science and Engineering, pp. 1–12 (2017, accepted)

# MRI-Based Computational Hemodynamics in Patients

**Andreas Ruopp and Ralf Schneider**

**Abstract** The target of this research was to develop a simulation process chain for the analysis of arterial hemodynamics in patients with automatic calibration of all boundary conditions for the physiological correct treatment of flow rates in transient blood flows with multiple bifurcations. The developed methodology uses stationary simulations at peak systolic acceleration and minimizes the error of target and simulated outflow conditions by means of a parallel genetic optimization approach. The target inflow and outflow conditions at peak systole are extracted from 4D phase contrast magnetic resonance imaging (4D PC-MRI). The flow resistance of the arterial system lying downstream of the simulation domain's outlets is modelled via porous media with velocity dependent loss coefficients. In the analysis of the subsequent transient simulations, it will be shown that the proposed calibration method shows to work suitable for three different types of patients including one healthy patient, a patient suffering from an aneurysm as well as one with a coarctation. Additionally the local effects of mapping the measured transient 4D PC-MRI data onto the aortic valve inlet in comparison to the usage of block inlet profiles will be shown.

## 1 Introduction

Computational fluid dynamics (CFD) can help to visualize and understand the flow behaviour in the human arterial system. Nowadays, patient specific CFD models are created from MRI or computer tomography (CT) in order to get realistic geometries and to evaluate pressure gradients in regions of coarctations [6, 9]. In future, CFD methods will help to reduce the need for diagnostic catheterization. In this context, new studies treat both pre- and post-treatment in CFD studies to improve their accuracy based on real patient data [5, 10]. Generally, 4D-MRI data are used to obtain the flow rates in the ascending and descending aorta. All other outlets in between are treated according to methods which rely on cross-sectional area relationships [5, 8]. The simulations have in common, that outflow conditions with

---

A. Ruopp (✉) • R. Schneider
HLRS, University of Stuttgart, Nobelstrasse 19, 70569 Stuttgart, Germany
e-mail: ruopp@hlrs.de; schneider@hlrs.de


M.M. Resch et al. (eds.), *Sustained Simulation Performance 2017*,
DOI 10.1007/978-3-319-66896-3_11

fixed flow rates are identical for both pre- and post-treatment setups even though the flow resistance in the arteries under consideration might be significantly altered by e.g. angioplasty.

This paper deals with the correct utilization of outflow conditions, when flow rates can be extracted from measurements via MRI. In this case, fluxes would be patient based and outflow conditions are calibrated to each single patient in order to use Dirichlet conditions for pressure at all outlets in conjunction with velocity dependent loss coefficients of porous media. By this approach the flow resistance of the arterial system lying downstream of each outlet is modelled independently from the flow conditions in the simulation domain.

## 2 Methodology

The following section gives an overview about the applied CFD codes, the complete optimization chain as well as the meshing, the domain mapping and the treatment of stationary and transient boundary conditions. The established work flow shows to be valid for different cases of hemodynamics in patients. Additionally, the use of porous domains clearly stabilizes the solution procedure and suppresses pressure reflections in the beginning of transient runs leading to higher possible Courant numbers for implicit solvers.

### *2.1 Geometries*

In this study, we use three different types of arterial systems originating just behind the aortic valve with different numbers of bifurcations, see Fig. 1. Case a was

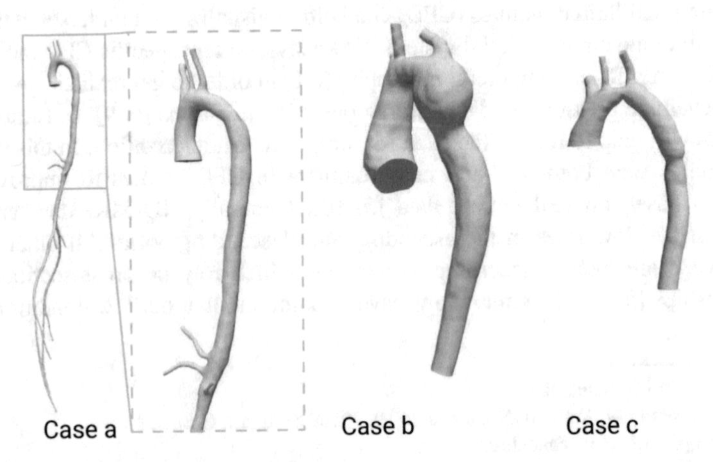


**Fig. 1** Three hemodynamic cases. a—Healthy patient, b—patient with aneurysm and c—patient with coarctation. All geometries in same scale except total view of a

extracted from CT while b and c are both extracted using PC-MRI data. Case b and c are inherited from Mirzaee et al. [10]. Case a consists of one inlet and thirteen outlets while case b and c have four outlets in total.

The edge resolution of the stereolithographic representation of case a and c is in the range of 0.0001 [m] for the most curved parts. Case b uses a higher resolution down to $10^{-6}$ [m]. All inlet and outlets are cut almost perpendicular with respect to the aortic wall in order to avoid numerical errors.

## 2.2 Meshing

We use a cartesian based meshing process *cartesianMesh* from cfMesh [2] for all three setups. One key advantage of this mesher is the fully automated workflow via scripting. Table 1 highlights the meshing parameters for each geometry setup. At each outlet, we introduce an additional volume mesh where the boundary mesh is extruded perpendicular to face normal's with ten layers each and an average thickness of about 0.0025 [m] per layer, see Fig. 2. The extrusion is done with the utility *extrudeMesh* which is included in the framework *OpenFOAM*®. In addition, each extruded mesh is marked with a *cellZone* being able to introduce specific source terms for porous media treatment during the simulation run.

**Table 1** Meshing parameters for *cfMesh*

| Parameter | Case | | |
|---|---|---|---|
| | a | b | c |
| Maximum cell size [m] | 0.001 | 0.001 | 0.001 |
| Refinement size of aortic wall [m] | 0.0005 | 0.001 | 0.0005 |
| Number of layers | 3 | 2 | 2 |
| Refinement of outlets [m] | 0.0005–0.00025 | 0.001 | 0.0005 |
| Cell numbers | 1,083,243 | 350,061 | 347,391 |

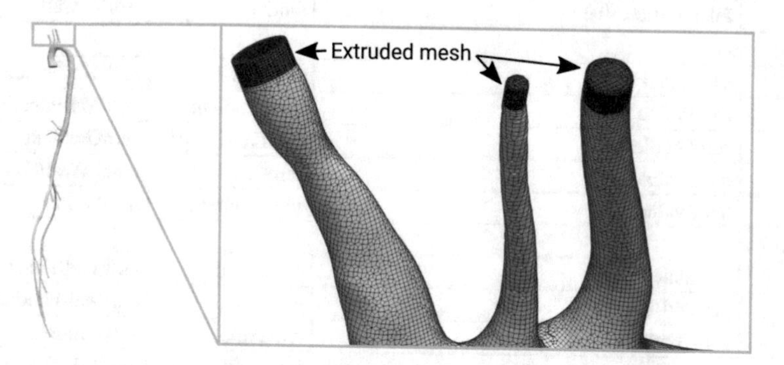


**Fig. 2** Extruded meshes at outlets for case a needed in conjunction with porous media. *Grey color* show mesh creation with *cfMesh*, *red color* indicates the extruded mesh afterwards

### 2.3 CFD Setup

Two CFD setups are performed including a steady state simulation setup for the optimization process followed by a transient case study. We assume, that the blood flow can be treated as in-compressible fluid behaviour. Therefore, we solve the steady and the unsteady Navier-Stokes equations in the following form

$$\frac{\partial \mathbf{U}}{\partial t} + (\mathbf{U} \cdot \nabla)\,\mathbf{U} - \nu \nabla^2 \mathbf{U} = -\frac{1}{\rho} \nabla p + S \tag{1}$$

$$\nabla \cdot \mathbf{U} = 0 \tag{2}$$

with $\nabla$ the Nabla operator, velocity $\mathbf{U}$, viscosity $\nu$, time $t$, pressure $p$, density $\rho$ and a source term $S$, using the solvers *simpleFoam* and *pimpleFoam* from *OpenFOAM*® Version 2.4.x respectively.

The porous media is introduced as a source term $S$, where we use an explicit porosity source *explicitPorositySource* with Darcy-Forchheimer model and the following relation

$$S = -\rho C_0 |\mathbf{U}|^{(C_1 - 1)} \mathbf{U} \tag{3}$$

$$C_0 \equiv \text{Model linear coefficient}$$

$$C_1 \equiv \text{Model Exponent coefficient}$$

with a fixed value of $C_1 = 2$ and a variable value of $C_0$ which has to be determined by optimization.

The boundary conditions are listed in Table 2 where $\nu_t$ denotes the turbulent viscosity of the fluid. We apply the *kOmegaSST*-model for all runs with a fixed

**Table 2** Boundary conditions for steady state and transient runs for all three cases

| | Inlet aortic valve | Outlet | Aortic wall |
|---|---|---|---|
| *Steady state* | | | |
| $\nu_t$ | calculated | calculated | nutkWallFunction |
| k | fixedValue | zeroGradient | kqRWallFunction |
| p | zeroGradient | fixedValue | zeroGradient |
| $\omega$ | fixedValue | zeroGradient | omegaWallFunction |
| **U** | fixedValue | zeroGradient | fixedValue |
| *Transient* | | | |
| $\nu_t$ | calculated | calculated | nutkWallFunction |
| k | groovyBC | groovyBC | kqRWallFunction |
| p | zeroGradient | fixedValue | zeroGradient |
| $\omega$ | groovyBC | groovyBC | omegaWallFunction |
| **U** | groovyBC (case a)<br>timeVaryingMappedFixedValue (case b,c) | zeroGradient | fixedValue |

kinematic viscosity of $0.004 \cdot 10^{-03}\,\mathrm{m}^2/\mathrm{s}$ and impose a fixed value for $k$ and $\omega$ at the aortic inlet with the following relations

$$k = 1.5 \cdot (|\mathbf{U}| \cdot I)^2 \tag{4}$$

$$I = 1\ [\%] \tag{5}$$

$$\omega = 0.09 \cdot \frac{\sqrt{k}}{r_{hydraulic}} \tag{6}$$

$$r_{hydraulic} = \sqrt{\frac{A_{Inlet}}{\pi}} \tag{7}$$

and a regular block profile for the velocity. For the transient cases, we apply a variable boundary condition that can switch between Dirichlet and Neumann dependent on the sign of the flux $\Phi_f$ at every boundary face. For $\Phi_f < 0$ the following values are set for $k$ and $\omega$ at each face

$$k = 1.5 \cdot (|\mathbf{U_f}| \cdot I)^2 \tag{8}$$

$$\omega = \max\left(0.09 \cdot \frac{\sqrt{k_f}}{r_{hydraulic}}\right) \tag{9}$$

where the subscript $f$ denotes the position of the considered face's centre. For $\Phi_f \geq 0$ we impose a zero gradient condition for $k$ and $\omega$ at each face.

The flow rate over time, given in Fig. 3, results in 5.1 $\left[\frac{1}{\text{min}}\right]$ for case a (inlet condition is similar to [11]), 5.35 $\left[\frac{1}{\text{min}}\right]$ for case b and case c leads to 4.51 $\left[\frac{1}{\text{min}}\right]$. For case b and c the measured velocity profiles, extracted from the according PC-MRI data were used at the inlet. The velocity vectors were interpolated in space and time onto the CFD mesh resulting in a more realistic inlet condition compared to

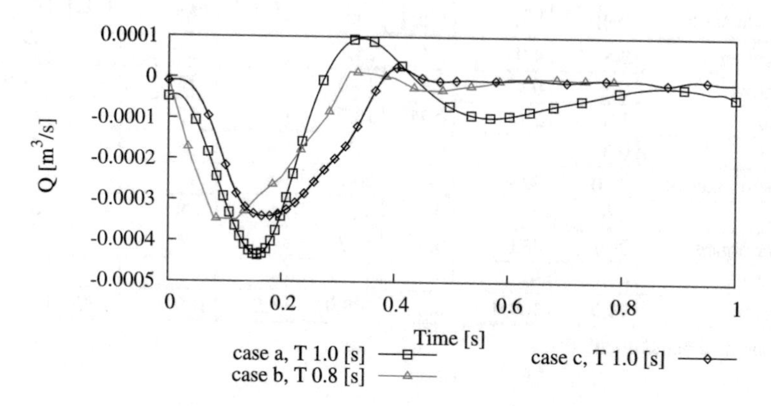


**Fig. 3** Given flow rate over time at inlet for case a, b and c

case a, see Fig. 4. Due to the turbulent velocity profile the flux dependent boundary condition at each cell face via *groovyBC* was needed.

No patient specific flow rates over time were available for case a at each outlet which is crucial for a correct calibration of entire hydrodynamic system. This also stands true for case b and c, since the evaluation of the 4D PC-MRI measurement data at each outlet violates the continuity equation with peak errors above 100 [%]. For this reason, we assume specific flow rates, derived from literature, which are summarized in Table 3. The target flow rates of all outlets for case a are given in Table 4 as well as for case b and c (Table 5) in terms of percentage of flow rate at inlet. We assume, that the given fractions stay constant during one heart beat.

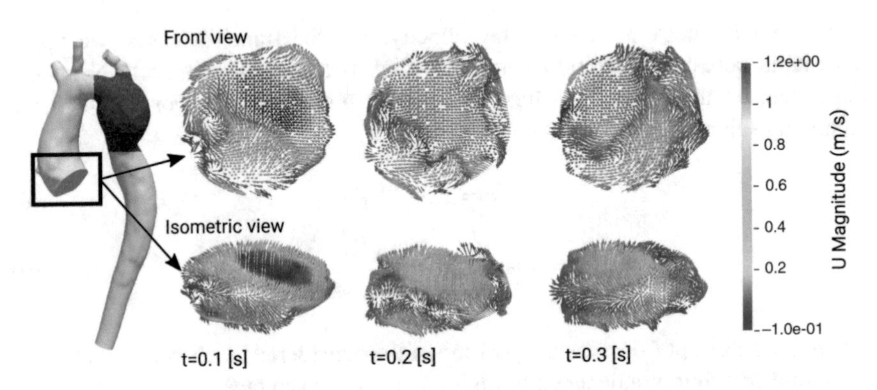


**Fig. 4** Velocity vectors at aortic inlet for case b in front view (*upper row*) and isometric view (*lower row*), *color* indicating magnitude of velocity. Aneurysm section colored *red* in geometry (*left picture*)

**Table 3** Division of volumetric flux at different bifurcations for case a, based on [13] (MiZ) and [7] (BWP)

| Classification | Weight [kg] | Fraction [%] | BWP $\left[\frac{1}{\min}\right]$ | Fraction [%] | MiZ $\left[\frac{1}{\min}\right]$ | Fraction [%] | $\left[\frac{1}{\min}\right]$ | Average [%] |
|---|---|---|---|---|---|---|---|---|
| Abdomen | 2.8 | 4.0 | 1.4 | 24.1 | – | 21 | – | 22.6 |
| Kidneys | 0.3 | 0.4 | 1.1 | 19.0 | – | 23 | – | 21.0 |
| Brain | 1.5 | 2.1 | 0.75 | 12.9 | – | 15 | – | 14.0 |
| Heart | 0.3 | 0.4 | 0.25 | 4.3 | – | 5 | – | 4.7 |
| Skeleton muscles | 30.0 | 42.9 | 1.2 | 20.7 | – | 17 | – | 18.8 |
| Skin | 5.0 | 7.1 | 0.5 | 8.6 | – | 8 | – | 8.3 |
| Other organs | 30.1 | 43.0 | 0.6 | 10.3 | – | 4 | – | 10.7 |
| Liver | – | – | – | – | – | 7 | – | – |
| Sum | 70.0 | 100.0 | 5.8 | 100.0 | 4.9 | 100 | 5.35 | 100.0 |

Missing data marked with –

**Table 4** Division of volumetric flux at different bifurcations for case a, based on [13] (MiZ) and [7] (BWP)

| | Flow rate $\left[\frac{l}{min}\right]$ | Fraction of inlet [–] |
|---|---|---|
| Inlet | 5.101 | 1.000 |
| Truncus Brachiocephalicus 03 | 0.549 | 0.108 |
| Arteria Carotis Communis 02 | 0.072 | 0.014 |
| Arteria Subclavia 01 | 0.331 | 0.065 |
| Truncus Coeliacus 04 | 1.202 | 0.236 |
| Arteria Mesentrica Superior 05 | 1.202 | 0.236 |
| Arteria Renalis Sinistra 06 | 0.561 | 0.110 |
| Arteria Renalis Dextra 07 | 0.561 | 0.110 |
| Arteria Iliaca Interna L 08 | 0.105 | 0.021 |
| Arteria Iliaca Interna R 09 | 0.105 | 0.021 |
| Arteria Profunda Femoris L 10 | 0.105 | 0.021 |
| Arteria Profunda Femoris R 11 | 0.105 | 0.021 |
| Arteria Femoralis L 12 | 0.102 | 0.020 |
| Arteria Femoralis R 13 | 0.102 | 0.020 |

**Table 5** Division of volumetric flux at different bifurcations for case b and c, based on [13] (MiZ) and [7] (BWP)

| | Flow rate $\left[\frac{l}{min}\right]$ | Frac. of inlet [–] | Flow rate $\left[\frac{l}{min}\right]$ | Frac. of inlet [–] |
|---|---|---|---|---|
| Inlet | 5.350 | 1.000 | 4.510 | 1.000 |
| Arteria Subclavia 01 | 0.327 | 0.061 | 0.275 | 0.061 |
| Arteria Carotis Communis 02 | 0.245 | 0.046 | 0.207 | 0.046 |
| Truncus Brachiocephalicus 03 | 0.381 | 0.071 | 0.321 | 0.071 |
| Outlet 04 | 4.397 | 0.822 | 3.707 | 0.822 |

## 2.4 *Optimization Workflow*

We apply an optimization workflow according to [12] for the steady state simulation runs in order to obtain the correct loss coefficient $C_0$ at each outlet. The optimization algorithm is identical to [4] which uses an evolutionary approach.

Case a uses in total 52 individuals per generation, case b and c 16 individuals. All loss coefficients were allowed to vary between $0.01–9 \cdot 10^6$. The optimization run strives for the minimum of the sum of each error between target flow rates and simulated flow rates

$$\epsilon = \sum_{i=1}^{n_{outlets}} \frac{\delta_i - \delta_{target,i}}{\delta_{target,i}} \tag{10}$$

# 3 Results

The following chapter shows the optimization results and the computational effort to calibrate systems with 13 or four unknowns respectively. Specific flow rates over time for different outlet sections and velocity distributions at different positions compared to the measured velocity field are presented for case b.

## 3.1 Optimization Results

In total, 10,000 designs had to be evaluated for case a, see Fig. 5, which takes 72 [h] using 1248 cores simultaneously on the HLRS Hazel Hen [3] system. Case b and case c, Fig. 6, with four outlets each need at least 300 and 1400 design evaluations respectively using 384 cores with a total time of at least 4 and 16 [h].

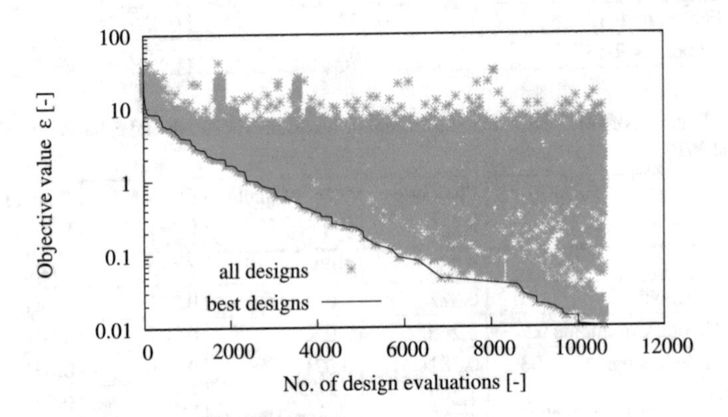


**Fig. 5** Objective value $\epsilon$ over number of simulation runs (individuals) for case a

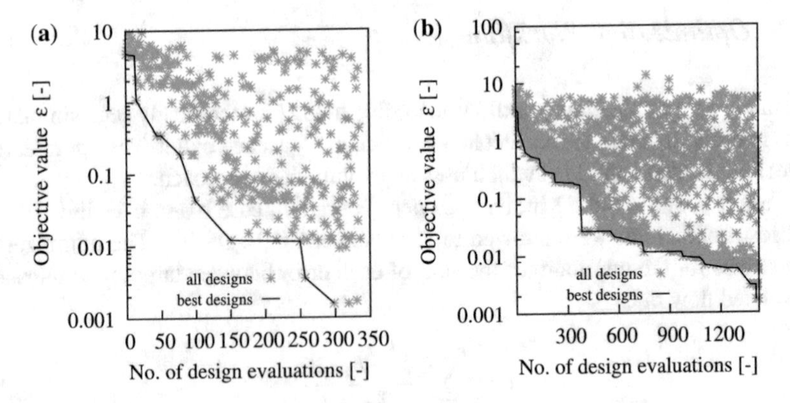


**Fig. 6** Objective value $\epsilon$ over number of simulation runs (individuals) for **(a)** case b and **(b)** case c

## 3.2 Global Target Values

The difference of volumetric fluxes over time for the calibrated and uncalibrated transient run for one heart beat is shown in Fig. 7. A calibration phase is necessary to obtain sensible fluxes over time. In addition, the presented porous media technique along with the shown calibration enables the simulation of one heart beat from rest. Without calibration and transient pressure boundary conditions at the outlets, one needs to simulate at least four heart beats to obtain a periodic transient state. This holds also true for case a and case b with a smaller number of outlets.

## 3.3 Local Variations

The local variations in magnitude of velocity are shown in Figs. 8, 9, and 10. The simulation run with calibrated outlets and correct mapped velocity field at the aortic inlet gives a relative good quantitative result. The uncalibrated run fails to capture

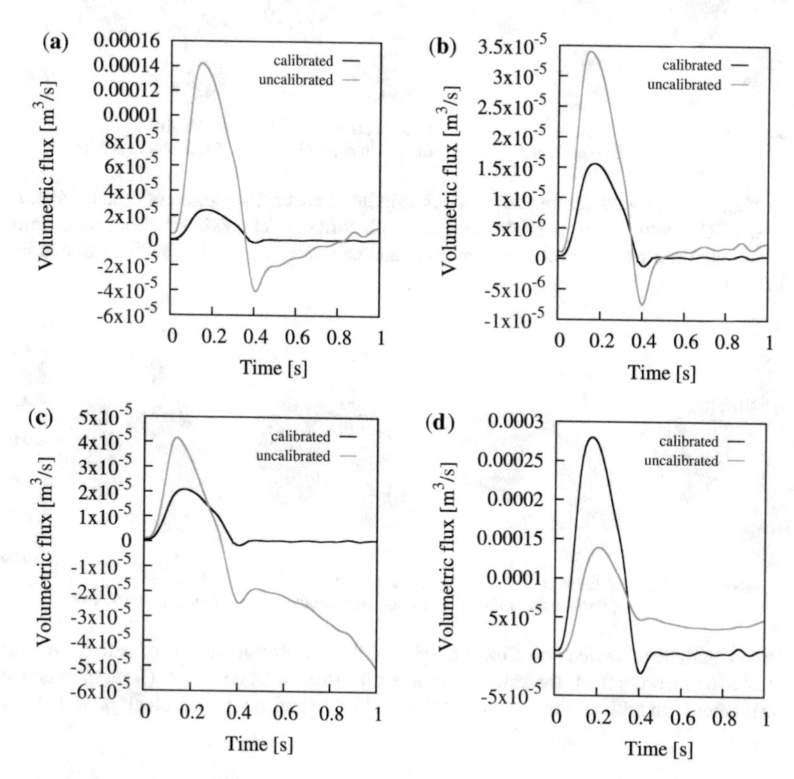


**Fig. 7** Volumetric fluxes over time at all four outlets for case c for first heart beat. (**a**) Outlet 01. (**b**) Outlet 02. (**c**) Outlet 03. (**d**) Outlet 04

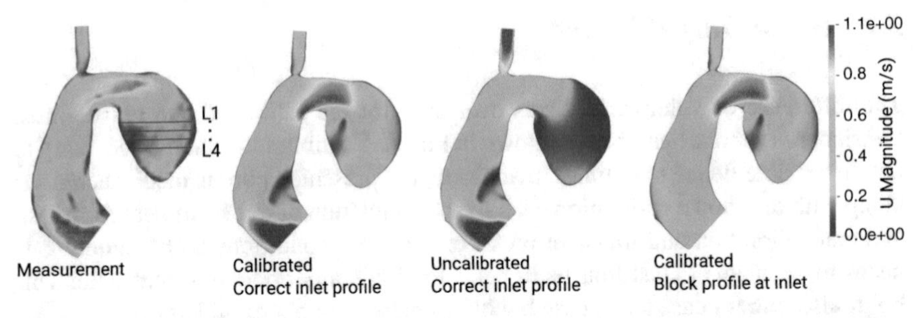


**Fig. 8** Magnitude of velocity on slice through middle of aneurysm section of case b. *From left to right*: PC-MRI measurement, the calibrated run with correct inlet condition (from measurement), the uncalibrated run with correct inlet condition and calibrated run with a block profile at inlet at $t = 0.12$ [s]. *Lines* (L1–L4) indicate probing position for quantitative comparison, see Appendix with Figs. 11 and 12

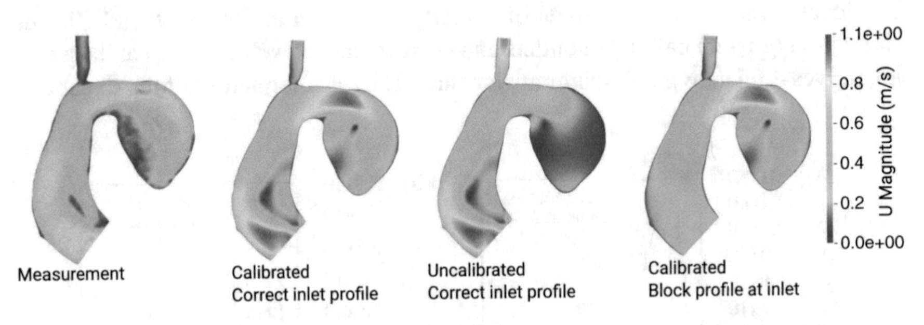


**Fig. 9** Magnitude of velocity on slice through middle of aneurysm section of case b. *From left to right*: PC-MRI measurement, the calibrated run with correct inlet condition (from measurement), the uncalibrated run with correct inlet condition and calibrated run with a block profile at inlet at $t = 0.16$ [s]

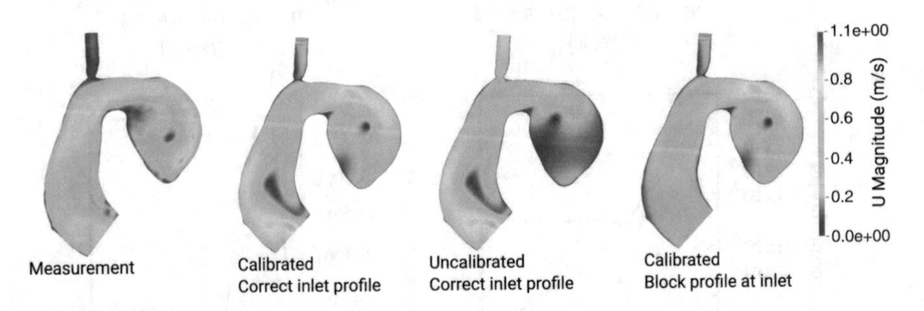


**Fig. 10** Magnitude of velocity on slice through middle of aneurysm section of case b. *From left to right*: PC-MRI measurement, the calibrated run with correct inlet condition (from measurement), the uncalibrated run with correct inlet condition and calibrated run with a block profile at inlet at $t = 0.20$ [s]

correctly the swirling flow field in the aneurysm section due to an unphysiological high volumetric flux in the region of the *Arteria Subclavia*. The calibrated run with the wrong inlet condition produces valid results behind the arch but the region next to the inlet is not captured correctly. In addition, the velocity distribution along four lines in the aneurysm section is shown in Figs. 11 and 12 again for the identical time steps. This quantitative comparison clearly shows a mismatch between the measurements and the simulation because the PC-MRI data (MRT) does have some inaccuracy according to each velocity component in dependency of the position. In addition, the simulation neglects fluid-structure interactions as well as detailed roughness estimations of aortic walls.

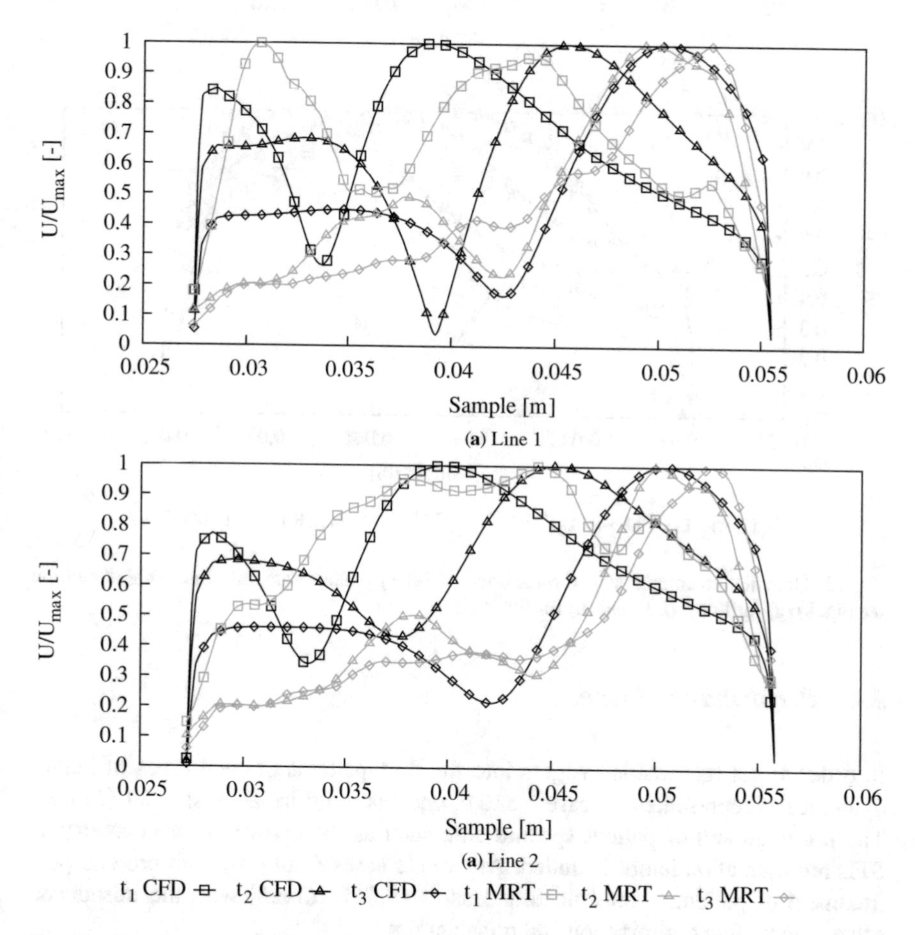


**Fig. 11** Quantitative comparison of magnitude of velocity along sampling lines (line definition see Fig. 8) (**a**) for line 1 (L1) and (**b**) for line 2 (L2)

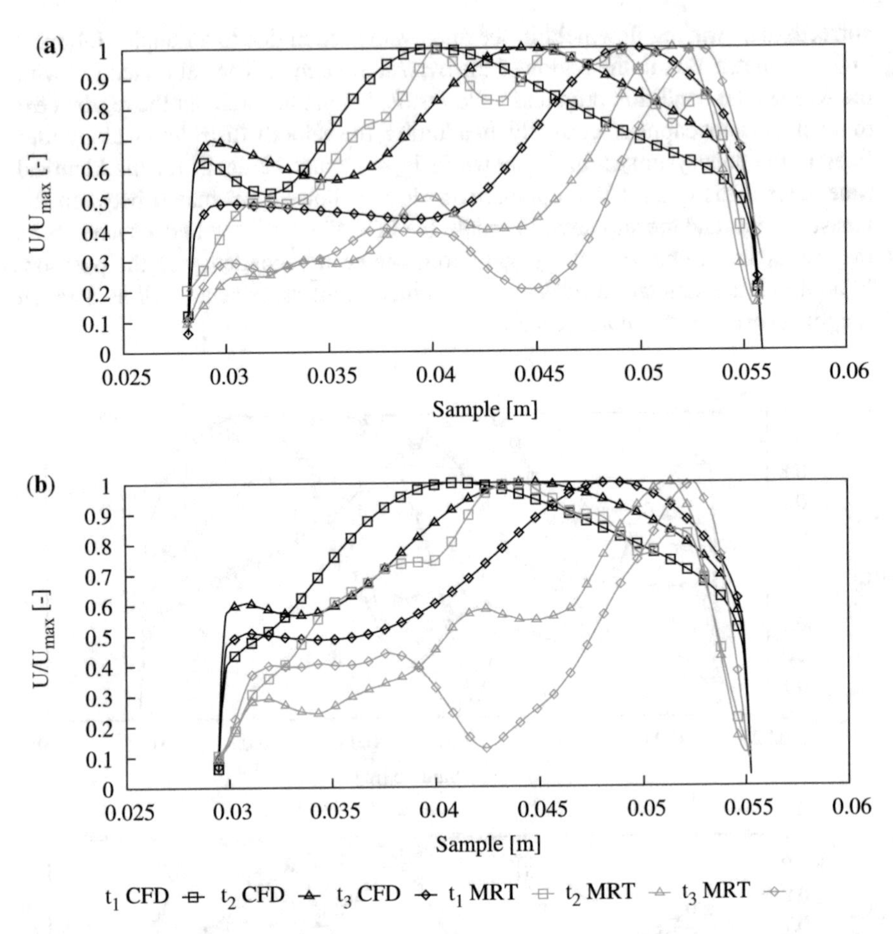


**Fig. 12** Quantitative comparison of magnitude of velocity along sampling lines (line definition see Fig. 8) (**a**) for line 3 (L3) and (**b**) for line 4 (L4)

## *3.4 Performance Issues*

In order to get reasonable insights into the flow patterns of such types of hemodynamics as represented by case b and c, one has to utilize at least 1560 [*Coreh*]. The pre-treatment of patient specific data such as the extraction of geometry as STL representation is not included. All twelve cases from [10] with pre and post treatment of patients would need at least ≈19,065 [*Coreh*] with the introduced scheme including optimization and transient run.

To estimate case counts that could be expected if the methodologies described above should be applied to indications found in typical cohort sizes, which are regarded as reasonable in classical clinical studies, we consider the cases described above. In average cohort sizes, we consider ≈5500 individuals (extracted from [1, 14]) with an average prevalence of 7.3 [%] of the regarded indication. This as a basis for further resource estimations of virtual clinical trials (see also [15]) leads to $\approx$400 individuals to which the simulation process has to be applied. If one takes into account one optimization to the preoperative state in conjunction with one preoperative and at least three postoperative transient evaluations in total, at least 656,000 [*Coreh*] are needed. This number of core-hours is the equivalent of 1 [day] facilitation of 1140 nodes on the HLRS Hazel Hen system [3].

## 4 Conclusion

Three patient specific geometries are simulated in a fully automated simulation process chain. The boundary conditions are treated with porous media with velocity dependent loss coefficients that are calibrated to physiological flow rates. By means of a parallel optimization process, aortic systems with up to 13 outlets can be calibrated in an adequate time. The transient simulation results clearly show the need of fully transient boundary conditions at the inlet, which have to be mapped from measurements in space and time. This enables a qualitative correct flow field in the complete domain in contrast to other assumptions such as block or parabolic profiles. At the moment, the lack of correct extraction of volumetric fluxes over time at each outlet for the target criteria is overcome by use of physical sensible estimations. Fixed pressure values at outlets in conjunction with the porous media model, even in varying conditions over time, can reproduce the correct flux balance.

In the sense of virtual clinical trials, an adequate number of individuals need to be investigated leading to a not insignificant usage of HPC systems. The presented estimation does not include fluid-structure interactions and non-Newtonian fluids.

## Appendix

See Figs. 11 and 12.

## References

1. Alcorn, H.G., Wolfson, S.K., Sutton- H., O'Leary, D.: Risk factors for abdominal aortic aneurysms in older adults enrolled in the cardiovascular health study. Arterioscler. Thromb. Vasc. Biol. **16**, 963–970 (1996)
2. cfMesh, http://cfmesh.com/, 14 Sept 2015
3. Cray XC40 (Hazel Hen). http://www.hlrs.de/systems/cray-xc40-hazel-hen/, May 2017
4. Deb, K., Tiwari, S.: Omni-optimizer: a generic evolutionary algorithm for single and multi-objective optimization. Eur. J. Oper. Res. **185**, 1062–1087 (2008)

5. Goubergrits, L., Riesenkampff, E., Yevtushenko, P., Schaller, J., Kertzscher, U., Hennemuth, A., Berger, F., Schubert, S., Kuehne, T.: MRI-based computational fluid dynamics for diagnosis and treatment prediction: clinical validation study in patients with coarctation of aorta. J. Magn. Reson. Imaging **41**, 909–916 (2015)
6. LaDisa, J.F., Alberto Figueroa, C., Vignon-Clementel, I.E., Jin Kim, H., Xiao, N., Ellwein, L.M., Chan, F.P., Feinstein, J.A., Taylor, C.A.: Computational simulations for aortic coarctation: representative results from a sampling of patients. J. Biomech. Eng. **133**, 091008–091008-9 (2011)
7. Lang, F., Lang, P.: Basiswissen Physiologie. 2nd edn. Springer, Berlin, Heidelberg (2007). http://dx.doi.org/10.1007/978-3-540-71402-6. ISBN:978-3-540-71402-6
8. Lantz, J., Karlsson, M.: Large eddy simulation of LDL surface concentration in a subject specific human aorta. J. Biomech. **45**, 537–542 (2012)
9. Menon, P.G., Pekkan, K., Madan, S.: Quantitative hemodynamic evaluation in children with coarctation of aorta: phase contrast cardiovascular MRI versus computational fluid dynamics. In: Statistical Atlases and Computational Models of the Heart. Imaging and Modelling Challenges: Third International Workshop, STACOM 2012, Held in Conjunction with MICCAI 2012, Nice, October 5 (2012). Revised Selected Papers. Camara, O., Mansi, T., Pop, M., Rhode, K., Sermesant, M. Young, A. (eds.), pp. 9–16. Springer, Berlin, Heidelberg (2013)
10. Mirzaee, H., Henn, T., Krause, M.J., Goubergrits, L., Schumann, C., Neugebauer, M., Kuehne, T., Preusser, T., Hennemuth, A.: MRI-based computational hemodynamics in patients with aortic coarctation using the lattice Boltzmann methods: clinical validation study. J. Magn. Reson. Imaging **45**, 139–146 (2017)
11. Patel, N., Küster, U.: Geometry dependent computational study of patient specific abdominal aortic aneurysm. In: Resch, M.M., Bez, W., Focht, E., Kobayashi, H., Patel, N. (eds.) Sustained Simulation Performance 2014. Proceedings of the Joint Workshop on Sustained Simulation Performance, University of Stuttgart (HLRS) and Tohoku University, pp. 221–238. Springer International Publishing, New York (2015)
12. Ruopp, A., Ruprecht, A., Riedelbauch, S.: Automatic blade optimisation of tidal current turbines using OpenFOAM®. In: 9th European Wave and Tidal Energy Conference (EWTEC), Southampton (2011)
13. Schaal, S., Steffen, K., Konrad, K.S.: Der Mensch in Zahlen: Eine Datensammlung in Tabellen mit über 20000 Einzelwerten, 4th edn. Springer Spektrum, Berlin (2016). http://dx.doi.org/10.1007/978-3-642-55399-8. ISBN:978-3-642-55399-8
14. Singh, K., Bønaa, K.H., Jacobsen, B.K., Bjørk, L., Solberg, S.: Prevalence of and risk factors for abdominal aortic aneurysms in a population-based study the Tromsø study. Am. J. Epidemiol. **154**, 236 (2001)
15. Viceconti, M., Henney, A., Morley-Fletcher, E.: In silico clinical trials: how computer simulation will transform the biomedical industry. Research and Technological Development Roadmap, Technical Report, 26 Mar 2017. http://dx.doi.org/10.13140/RG.2.1.2756.6164

# Part V
# High Performance Data Analytics

# A Data Analytics Pipeline for Smart Healthcare Applications

Chonho Lee, Seiya Murata, Kobo Ishigaki, and Susumu Date

**Abstract** The rapidly increasing availability of healthcare data is becoming the driving force for the adoption of data-driven approaches. However, due to a large amount of heterogeneous dataset including images (MRI, X-ray), texts (doctor's note) and sounds, doctors still struggle against temporal and accuracy limitations when processing and analyzing such big data using conventional machines and approaches. Employing advanced machine learning techniques on big healthcare data anlaytics supported by Petascale high performance computing resources is expected to remove those limitations and help find unseen healthcare insights. This paper introduces a data analytics pipeline consisting of data curation (including cleansing, annotation, and integration) and data analytics processes, necessary to develop smart healthcare applications. In order to show its practical use, we present sample applications such as diagnostic imaging, landmark extraction and casenote generation using deep learning models, for orthodontic treatments in dentistry. Eventually, we will build smart healthcare infrastructure and system that fully automate the set of the curation and analytics processes. The developed system will dramatically reduce doctor's workload and is smoothly expanded to other fields.

## 1 Introduction

Increasing demand and costs for healthcare, exacerbated by ageing populations, are serious concerns worldwide. A relative shortage of doctors or clinical manpower is also a big problem that causes their workload to increase and brings a challenge for them to provide immediate and accurate diagnoses and treatments for patients. Most of the medical practices are completed by medical experts backed by their own experiences, and clinical researches are conducted by researchers via painstaking

C. Lee (✉) • S. Date
Cybermedia Center, Osaka University, 5-1 Mihogaoka, Ibaraki, Osaka, Japan
e-mail: leech@cmc.osaka-u.ac.jp; date@cmc.osaka-u.ac.jp

S. Murata • K. Ishigaki
Graduate School of Information Science and Technology, Osaka University, 5-1 Mihogaoka, Ibaraki, Osaka, Japan
e-mail: murata.seiya@ais.cmc.osaka-u.ac.jp


M.M. Resch et al. (eds.), *Sustained Simulation Performance 2017*,
DOI 10.1007/978-3-319-66896-3_12

designed and costly experiments. Consequently, this has generated a great amount of interests and motivation in providing better healthcare through smarter healthcare systems.

Nowadays, a huge amount of healthcare data, called Electronic Health Records (EHR), has become available in various healthcare organizations, which are the fundamental resource to support medical practices or help derive healthcare insights. The increasing availability of EHR is becoming the driving force for the adoption of data-driven approaches. Efficient big healthcare data analytics supported by advanced machine learning (ML) and high performance computing (HPC) technologies brings the opportunities to automate healthcare related tasks. The benefits may include earlier disease detection, more accurate prognosis, faster clinical research advance and the best fit for patient management.

While the promise of big healthcare data analytics is materializing, there is still a non-negligible gap between its potential and usability in practice due to various factors that are inherent in the data itself such as high-dimensionality, heterogeneity, irregularity, sparsity and privacy. To make the best analytics, all the information must be collected, cleaned, integrated, stored, analyzed and interpreted in a suitable manner. The whole process is a data analytics pipeline where different algorithms or systems focus on different specific targets and are coupled together to deliver an end-to-end solution. It can also be viewed as a software stack where at each phase there are multiple solutions and the actual choice depends on the data type (e.g. image, sound, text, or sensor data) or application requirements (e.g. predictive analysis or cohort analysis).

In this paper, we describe the data analytics pipeline consisting of *data curation phase* with cleansing, annotation and integration, and *data analytics phase* with analytics methods and visualization tools, which are necessary processes to develop healthcare applications. In order to show its practical use, we present three example applications using deep learning methods for orthodontic treatments in dentistry. The applications try to automate the following tasks such as (1) computing Index of Orthodontic Treatment Needs (IOTN)[1] from facial and oral photos; (2) extracting facial morphological landmarks or features from X-rays called Cephalograms; and (3) generating casenote where the first doctor's observation is written based on the diagnostic imaging such as (1) and (2).

The remainder of this paper is organized as follows. Section 2 introduces some requirements in handling healthcare data and describes the proposed data analytics pipeline consisting of data curation and analytics processes. Section 3 presents example applications using deep learning models, such as diagnostic imaging, landmark extraction and casenote generation for orthodontic treatments in dentistry, followed by conclusion.

[1]IOTN [1] is one of the severity measures for malocclusion and jaw abnormality, which determines whether orthodontic treatment is necessary.

## 2 Healthcare Data Analytics Pipeline

This section describes the proposed healthcare data analytics pipeline. As illustrated in Fig. 1, it consists of two phases, *data curation phase* (Sect. 2.1) and *data analytics phase* (Sect. 2.2). Processing big data is supported by high performance computing resources.

### 2.1 Data Curation Phase

Before available data is directly processed for analysis, data needs to go through several steps to refine it according to application requirements. Data curation phase prepares necessary data in a suitable format for further analysis. Firstly, data needs to be acquired and extracted from various data sources. Secondly, obtained raw data is probably heterogeneous, composed of structured, unstructured and sensor data, and also typically noisy due to inaccuracies, missing, biased evaluations, etc.

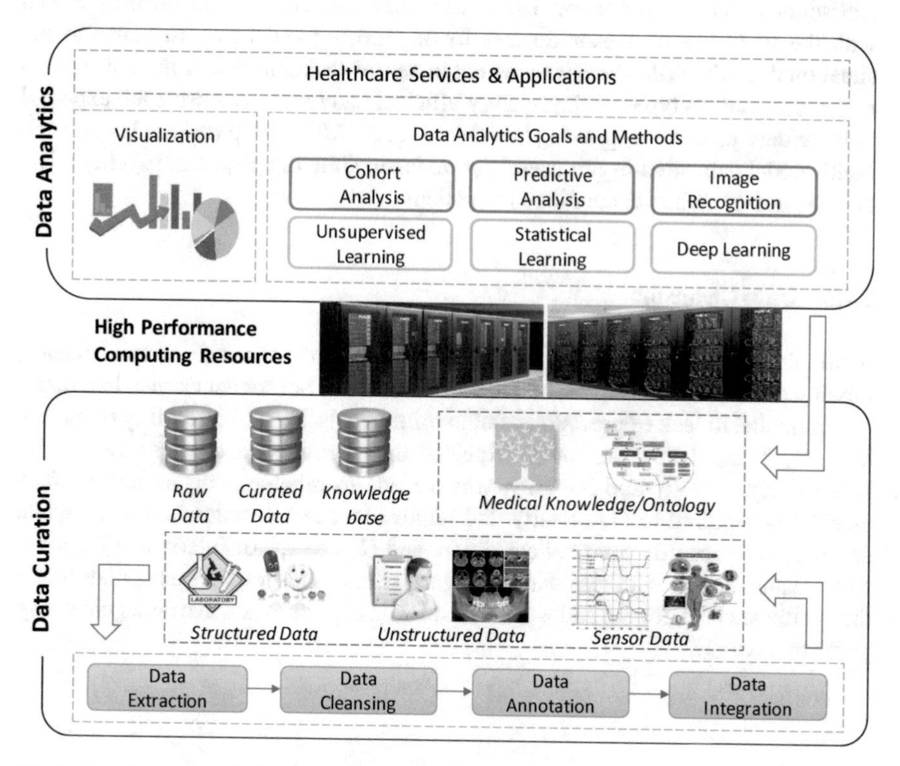


**Fig. 1** An illustration of the proposed data analytics pipeline consisting of data curation and analytics, necessary processes to develop healthcare applications

Hence, data cleansing is required to remove data inconsistencies and errors. Thirdly, data annotation with medical experts' assistance contributes to the effectiveness and efficiency of this whole process. Fourthly, data integration combines various sources of data to enrich information for further analysis. Finally, the processed data is modelled and analyzed, and then analytics results are visualized and interpreted.

### 2.1.1 Data Type

Healthcare data, e.g., electric healthcare records (EHR), mainly includes three types of data, namely structured, unstructured and sensor data. *Structured data* includes socio-demographic information and medical features such as diagnoses, lab test results, medications and procedures. Those elements are typically coded in pre-defined forms by a hierarchical medical classification system or IDC-9 and currently IDC-10,[2] and drug databases like First Databank and SNOMED (Systematic Nomenclature of Medicine). *Unstructured data* does not have a specific data model, which includes medical status in a free-text form (e.g., doctors' notes and medical certificates) and non-textual form such as images (e.g., MRI, X-rays) and sounds. *Sensor signals* or data streams are also common in healthcare data with the wide use of sensor devices for monitoring and better response to the situational needs. With the advancement in sensor technology and miniaturization of devices, various types of tiny, energy-efficient and low-cost sensors are expected to be widely used for improving healthcare [2, 3]. Monitoring and analyzing such multi-modal data streams are useful for understanding the physical, psychological and physiological health conditions of patients.

### 2.1.2 Data Cleansing

Available raw data is typically noisy due to several reasons such as inaccuracies, missing data, erroneous inputs, biased evaluations, etc. Sensor data is also inherently uncertain due to lack of precisions, failure of transmission and instability of battery life, etc. Thus, data cleansing is expected to improve data quality assessed by its accuracy, validity and integrity, which leads to reliable analysis results. It is essentially required to (1) identify and remove inaccurate, redundant, incomplete and irrelevant records from collected data and (2) replace or interpolate incorrect and missing records with reasonably assigned values. This requires us to understand the healthcare background and work with domain experts to achieve better cleansing performance.

[2]International Statistical Classification of Diseases and Related Health Problems.

#### 2.1.3 Data Annotation

Incompleteness is a common issue in terms of data quality. Although the uncertainty of data can be resolved by model inference using various learning techniques, most healthcare data is inherently too complex to be inferred by machines using limited information. In such cases, enriching and annotating data by medical experts are the only choice to help the machine to correctly interpret data. However, the acquisition of supervised, annotated information results in an expensive exploitation of data.

Active learning is one of the approaches to reduce the annotation cost while learning algorithms achieve higher accuracy with few labelled training data. It aims to only annotate the important, informative data instances while inferring others, and thereby the total number of annotated data is significantly reduced. The general solutions may include reducing the uncertainty of training models by uncertainty sampling [4], Query-By-Committee [5], maximizing the information density among the whole query space [6]. Another approach may be to borrow knowledge from related domain(s) such as transfer learning [7]. However, the aforementioned methods have limitations in real healthcare applications due to healthcare data volume, complexity and heterogeneity. The automation of data annotation is still a challenging problem.

#### 2.1.4 Data Integration

Data integration is the process of combining heterogeneous data from multiple sources to provide users with a unified view of these data. Gomez et al. [8] explores the progress made by the data integration community, and Doan [9] introduces some principles as well as theoretical issues in data integration.

Typically, EHR integrates heterogeneous data from different sources including structured data such as diagnoses, lab tests, medications, unstructured free-text data like discharge summary, image data like MRI, etc. Healthcare sensor data is generated by various types of sensor/mobile devices at different sampling rates. The heterogeneity of abundant data types brings another challenge when integrating data streams due to a tradeoff between the data processing speed and the quality of data analytics. The high degree of multi-modality increases the reliability of analytics results, but it requires longer data processing time. The lower degree of multi-modality will improve data processing speed but degrade the interpretability of data analytics results. The efficient data integration helps reduce the size of data to be analyzed without dropping the analysis performance (e.g., accuracy).

### *2.2 Data Analytics Phase*

Data anlaytics phase (the upper box of Fig. 1) applies different analytics methods into the curated data to retrieve medical knowledge. Visualization techniques may also be used to get better understanding of the data. Utilizing high performance computing resources, we can improve the efficiency of data analysis especially when dealing with a large scale of data. For data privacy, on-demand secure network connection will be established, in which data is located or transferred to compute nodes when only needed. Right after the computation, the connection will be disconnected, and the data will be deleted.

### 2.2.1 Analytics Methods

Among a variety of anlaytics methods, the actual choice of algorithms or solutions depends on the data type (e.g. image, sound, text, sensor data) and/or application requirements (e.g. cohort analysis, predictive analysis, image recognition). In this section, we shall introduce a few basic methods to solve some healthcare problems as shown in Fig. 1.

- *Cohort Analysis*: Cohort analysis is a technique to find risk factors in a particular group of people, who have certain attributes or conditions such as birth, living area, life style, medical records, etc. The group is compared with another group who are not affected by the conditions. Long term statistical investigation will assess the significant differences between them. For the cohort analysis, clustering or unsupervised learning is the most popular method to divide people into particular groups under the certain conditions. For example, Sewitch [10] identifies multivariate patterns of perceptions using clustering method. Five different patient clusters are finally identified and statistically significant inter-cluster differences are found in psychological distress, social support satisfaction and medication non-adherence.
- *Predictive Analysis*: Disease progression modeling (DPM) is one of the predictive analysis, which employs computational methods to model the progression of a specific disease [11]. Reasonable prediction using DPM can effectively delay patients' deterioration and improve their healthcare outcomes. Typically, statistical learning methods are applied to find a predictive function based on historical data, i.e., the correlation between medical features and condition indicators. For example, Schulze [12] uses a multivariate Cox regression model that computes the probability of developing diabetes within 5 years based on anthropometric, dietary, and lifestyle factors.
- *Image Recognition*: Analyzing medical images such as X-ray, MRI, etc. are beneficial for many medical diagnosis and a wide range of the studies focus on classification or segmentation tasks. The recent breakthrough in image recognition technology using deep convolutional neural network (CNN) model [13] brings further improvement in diagnostic imaging that can diagnose the presence of tuberculosis in chest X-ray images [14], detect diabetic retinopathy from retinal photographs [15], as well as locate breast cancer in pathology images [16]. A model that combines deep learning algorithms and deformable models is developed in [17] for fully automatic segmentation of the left ventricle from cardiac MRI datasets.

Although a variety of data analytics and machine learning (ML) tools are available, there still exists an obstacle for doctors to fully utilize the tools due to the lack of the usability. Besides, it is difficult for them to manage compute resources suitable for executing the analytics methods. Hence, in near future, high performance infrastructure and system that operate fully or semi-fully automated big healthcare data curation and analytics, are eagerly desired in medical environment so that any doctors and/or researchers efficiently conduct their own data analytics.

## 3 Smart Orthodontic Treatment in Dentistry

The recent breakthrough in image recognition using deep learning techniques brings further improvement in diagnostic imaging. The diagnostic imaging is eagerly desired in the field of orthodontics as well, along with increasing demands for dental healthcare, becoming one of the regular life health factors. For example, the remote diagnostic imaging can evaluate malocclusion and jaw abnormality that are the causes of masticatory dysfunction, apnea syndrome and pyorrhea, etc.

In orthodontic clinic, a patient is generally taken his/her facial and oral images from all directions (as shown in Fig. 2) and given the first observation. Looking at the images, doctors spend time discussing the observation and create the medical

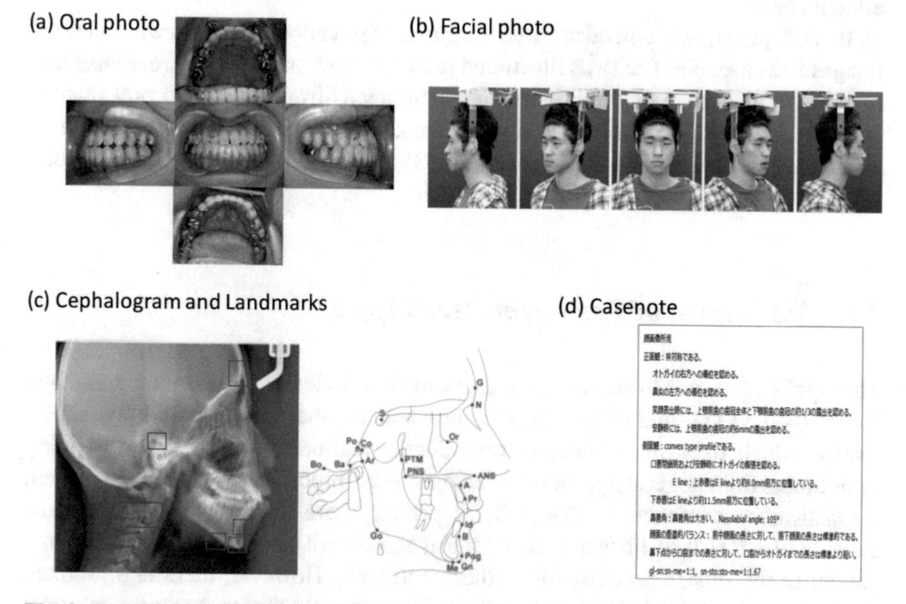


**Fig. 2** Sample dataset collected for orthodontic treatments. (**a**) Oral images taken from five directions, (**b**) facial images taken from five directions, (**c**) Cephalogram [18], Morphological landmarks [19] and example patch images (*red boxes*) (**d**) casenote or the first doctor's observation

records including diagnosis, treatment plan and progress checkup, etc. This process is certainly necessary for providing objective diagnosis that is important for both doctors and patients because the diagnosis directly affects to the treatment plan, treatment priority, and insurance coverage; but, it takes a great deal of time. In Osaka University Dental Hospital, over thousands of patient visits are counted including about 100 new patients per year. It is overburdened for doctors to properly manage a sequence of tasks such as diagnosis, treatment, progress checkup and counselling for all patients. Especially, doctors spend a lot of time and effort to diagnose by manually looking at massive number of images; for instance, it takes about 2–3 h for just one patient's case. The automation of diagnostic imaging is highly expected to assist doctors reducing their workload and providing objective diagnosis.

We try to develop a high performance infrastructure that operates big healthcare data analytics systems, especially for orthodontic treatments in dentistry, which automate medical tasks such as diagnostic imaging, landmark extraction and casenote generation. Due to a large amount of heterogeneous dataset including images (facial/oral photo, X-rays) and texts (casenote), doctors struggle against temporal and accuracy limitations when processing and analyzing those data using conventional machines and approaches. We believe that advanced machine learning techniques supported by Petascale high performance computing infrastructure remove those limitations and help find unseen healthcare insights. We evaluate the practical use of DL models in medical front and show its effectiveness.

In this paper, we consider three example applications dealing with medical images and casenotes (text), as illustrated in Fig. 3. Section 3.1 (App1) explains how to compute the score of orthodontic treatment needs from facial and oral images. Section 3.2 (App2) shows how to retrieve facial morphological landmarks from X-rays called Cephalograms. Section 3.3 (App3) describes how to generate casenotes where the first doctor's observation is written.

### *3.1 Assessment of Treatment Need (App 1)*

This application tries to automate the assessment of Index of Orthodontic Treatment Needs (IOTN) [1], one of the severity measures for malocclusion and jaw abnormality, which determines whether orthodontic treatment is necessary. Providing orthodontic treatments at appropriate timing is very important for patients to prevent a masticatory dysfunction. Generally, a primary care doctor or general dentist assesses the IOTN of his/her patient, and if the severity is high, he/she refers the patient to the other specialist for further treatments. However, there is a problem that many patients tends to miss the appropriate treatment timing due to an incorrect assessment by an inexperienced doctor. The automation of the IOTN assessment helps provide an objective assessment and train such inexperienced doctors.

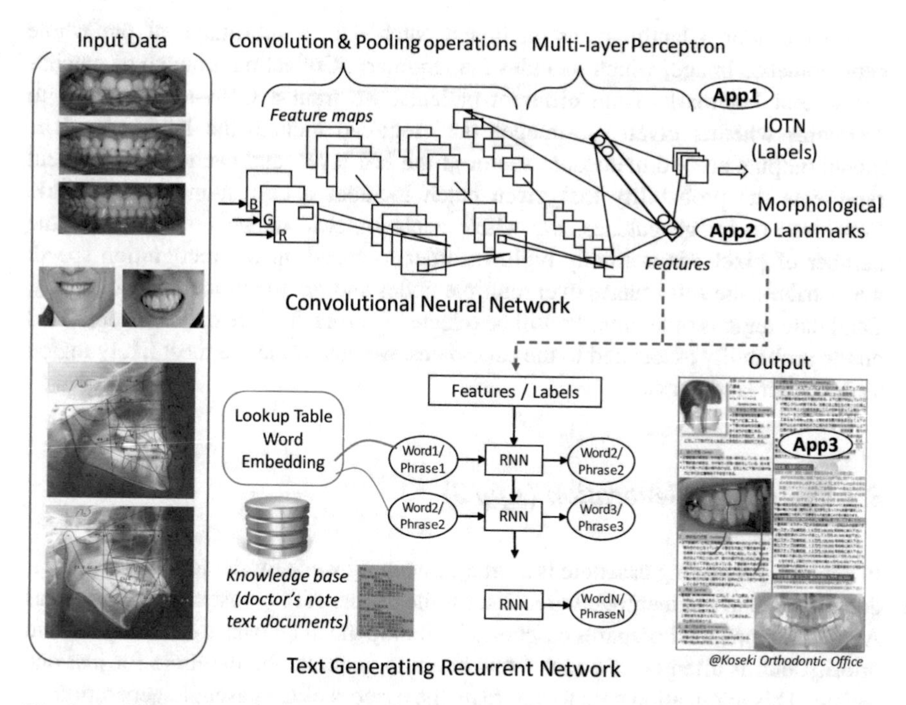


**Fig. 3** An illustration of deep learning models that perform diagnostic imaging (App1), landmark extraction (App2) and casenote generation (App3) for orthodontic treatments

We collect oral and facial images of over a thousand patients, taken from five different directions, as shown in Fig. 2a, b. Unlike typical image classification problems where each image is paired with one class or label, one class, i.e., a severity value, is paired with a set of images of a patient. We design a parallel convolutional neural network (CNN) model that independently runs multiple CNNs, each of which deals with images taken from each direction, and then concatenates feature vectors (i.e., outputs of the multiple CNNs). The concatenated feature vector is input to a multi-layer perceptron whose output is one of IOTN levels.

## 3.2 *Morphological Landmarking (App 2)*

Cephalometric analysis is also a significant diagnosis necessary for further orthodontic treatments. It helps in classification of skeletal and dental abnormalities, planning treatment of an individual, and predicting growth related changes. This application tries to automate facial morphological landmark detection in Cephalometric X-ray images (Fig. 2c).

We consider a landmark as an image patch, i.e., a sub-image of the whole cephalometric image, which includes the landmark. Collecting a bunch of patches for several landmarks from different patients, we train a CNN-based model to recognize whether given sub-images (i.e., regions) include the landmarks. The model outputs an $N$ dimensional vector at the last layer, and each vector element represents the probability that given patch includes a corresponding landmark. Compared to image patches, the whole cephalometric image resolution (or the number of pixels) is normally high. In order to speed up the recognition speed, we distribute the sub-images over multiple nodes and run the model independently. Candidate regions of landmarks will be selected from each of the nodes. Then, based on the probability associated to the candidates, we determine the most likely region as the target landmark.

### *3.3 Casenote Generation (App 3)*

In general, generating casenote is a time consuming work for doctors. For instance, doctors regularly gather together to discuss the results of diagnostic imaging such as App1 and App2, and prepares casenotes including the diagnosis, treatment plan and priority, etc. It often takes about a few hours to generate the casenote for just one patient. This application tries to automate the process of the casenote generation.

We collect over a thousand of casenotes (Fig. 2d) in addition to the oral and facial images of the corresponding patients. Inspired by a related work [20] that describes the content of an image, we design a hybrid model using CNN and Recurrent neural network (RNN) that inputs both images and casenotes. The hybrid model will lean how the casenote was written according to the diagnostic imaging; in other words, the model will find the association rules between features in images and words in casenotes.

## 4 Conclusion

In this paper, we summarize some requirements in handling healthcare data and the data analytics. We propose a data analytics pipeline that consists of data curation with cleansing, annotation and integration, and data analytics processes using several analytics methods and visualization tools. In order to verify the practical use of such data curation and analytics methods in medical front and show its effectiveness, we present example healthcare applications such as diagnostic imaging, landmark extraction and casenote generation using deep learning models, for orthodontic treatments in dentistry.

In future work, we will conduct rigorous experiments to evaluate the results of the curation and analytics and whether they satisfy the application requirements. Eventually, we will build smart healthcare infrastructure and system that fully or semi-fully automate the set of the curation and analytics processes, where any doctors and/or researchers efficiently conduct their own data analytics, which dramatically reduces their workload. This will be smoothly expanded to other fields such as otolaryngology (ear and nose) and ophthalmology (eye).

**Acknowledgements** The authors would like to thank Prof. Kazunori Nozaki in Osaka University Dental Hospital, for managing and providing medical dataset for experiments. We also thank Prof. Chihiro Tanikawa in Department of Orthodontics & Dentofacial Orthopedics, Osaka University Dental Hospital, for lending her expertise on the orthodontic treatments in dentistry.

## References

1. Brook, P.H., Shaw, W.C.: The development of an index of orthodontic treatment priority. Eur. J. Orthod. **11**(3), 309–320 (1989)
2. Caytiles, R.D., Park, S.: A study of the design of wireless medical sensor network based u-healthcare system. Int. J. Bio-Sci. Bio-Technol. **6**(3), 91–96 (2014)
3. Filipe, L., Fdez-Riverola, F., Costa, N., et al.: Wireless body area networks for healthcare applications. Protocol stack review. Int. J. Distrib. Sens. Netw. (2015). http://dx.doi.org/10.1155/2015/213705
4. Sharma, M., Bilgic, M.: Evidence-based uncertainty sampling for active learning. Data Min. Knowl. Discov. **31**(1), 164–202 (2017)
5. Seung, H.S., Opper, M., Sompolinsky, H.: Query by committee. In: Proceedings of the Fifth Annual Workshop on Computational Learning Theory (1992)
6. Settles, B., Craven, M.: An analysis of active learning strategies for sequence labeling tasks. In: Proceedings of the Conference on Empirical Methods in Natural Language Processing, EMNLP08 (2008)
7. Ma, Z., Yang, Y., Nie, F., Sebe, N., Yan, S., Hauptmann, A.: Harnessing lab knowledge for real-world action recognition. Int. J. Comput. Vis. **109**(1–2), 60–73 (2014)
8. Gomez-Cabrero, D., Abugessaisa, I., Maier, D., et al.: Data integration in the era of omics: current and future challenges. BMC Syst. Biol. **8**(2) (2014). http://dx.doi.org/10.1186/1752-0509-8-S2-I1
9. Doan, A., Halevy, A., Ives, Z.: Principles of Data Integration. Elsevier, Amsterdam (2012)
10. Sewitch, M.J., Leffondré, K., Dobkin, P.L.: Clustering patients according to health perceptions: relationships to psychosocial characteristics and medication nonadherence. J. Psychosom. Res. **56**(3), 323–332 (2004)
11. Mould, D.: Models for disease progression: new approaches and uses. Clin. Pharmacol. Ther. **92**(1), 125–131 (2012)
12. Schulze, M.B., Hoffmann, K., Boeing, H., et al.: An accurate risk score based on anthropometric, dietary, and lifestyle factors to predict the development of type 2 diabetes. Diabetes Care **30**(3), e89 (2007)
13. Krizhevsky, A., Sutskever, I., Hinton, G.E. : Imagenet classification with deep convolutional neural networks. In: Advances in Neural Information Processing Systems, vol. 1 (2012)
14. Lakhani, P., Sundaram, B.: Deep learning at chest radiography: automated classification of pulmonary tuberculosis by using convolutional neural networks. Radiology (2017). http://dx.doi.org/10.1148/radiol.2017162326
15. Gulshan, V., Peng, L., Coram, M., et al.: Development and validation of a deep learning algorithm for detection of diabetic retinopathy in retinal fundus photographs. J. Am. Med. Assoc. **316**(22), 2402–2410 (2016)
16. Janowczyk, A., Madabhushi, A.: Deep learning for digital pathology image analysis: a comprehensive tutorial with selected use cases. J. Pathol. Inf. **7**(292), 29 (2016)

17. Avendi, M.R., Kheradvar, A., Jafarkhani, H.: A combined deep-learning and deformable-model approach to fully automatic segmentation of the left ventricle in cardiac MRI. Med. Image Anal. **30**, 108–119 (2016)
18. Jimbocho Orthodontic clinic: www.jimbocho-ortho.com. Accessed June 2017
19. Grau, V., Alcaniz, M., Juan, M., Knoll, C.: Automatic localization of cephalometric landmarks. J. Biomed. Inform. **34**(3), 146–156 (2001)
20. Vinyals, O., Toshev, A., Bengio, S., Erhan, D.: Show and tell: a neural image caption generator. In: CVPR15 (2015)

Printed in the United States
By Bookmasters